aircraft illustrated

CIVIL AVIATION REVIEW

Edited by Leo Marriott

LONDON

IAN ALLAN LTD

Contents

First published 1989

ISBN 0 7110 1863 4

Published by Ian Allan Ltd, Shepperton, Surrey; and printed by Ian Allan Printing Ltd at their works at Coombelands in Runnymede, England

Front cover:
Delta Airlines Boeing 757 at Atlanta, Georgia in 1985.
Aviation Picture Library

Right:
An Embraer Brasilia operating on a Continental Express feeder service.
Embraer

Introduction

WELCOME to this the second edition of *Civil Aviation Review* which presents a wide ranging picture of the state of civil air operations around the world. Once again the amazing versatility of the aeroplane in serving the peaceful needs of the world's community is well illustrated in these pages.

Regrettably the past year has shown the down side of civil aviation with two major accidents in the United Kingdom as well as others around the world. It is a sad trick of fate that this issue should contain an article commemorating the first flight of the mighty Boeing 747, 20 years ago in 1969, but also featuring the disturbing details of the Lockerbie tragedy when one of these fine aircraft was brought down by the incomprehensible actions of a terrorist group. The result of this accident will, no doubt, mean even stricter security precautions at airports but the suspicion remains that determined and deviously cunning fanatics will somehow always find a way of circumventing any precautions. It is no exaggeration to say that a state of war exists between the airline industry and the world's terrorist groups, whoever they may be, and in any war the advantage will swing between the attack and defence. Regrettable though it may be, airline passengers must be prepared to put up with more and more restrictions in order to ensure their continuing safety.

The outcome of investigations of other accidents, increasingly discussed openly in all media channels, will also have contributed to a perceived shortfall in air safety. Statistically this may not be justified but civil aviation is perhaps a victim of its own good safety record and public expectations are higher. The trend towards larger aircraft means that the number of people involved in each accident is tending to increase so that each disaster seems correspondingly worse than those which have happened in the past. Like the battle against terrorism, a perfect safety record is probably unobtainable — but that will not stop the aviation industry from devoting its efforts to attaining such a goal.

The other major problem facing the air transport industry today is congestion in the air. Again, this is a product of success. As more and more people want to fly the numbers of aircraft operating in finite volumes of airspace has increased relentlessly. In this edition we look at some of the technical developments in air traffic control radar equipment which will provide controllers with new tools in the future. However, air traffic congestion cannot be solved overnight and requires new staff to be trained, new procedures for efficient use of airspace to be devised, and international agreements between governments on divisions of responsibility. While this continuing process is being implemented, the growth of airline traffic will unfortunately leave passengers sitting out long delays at airports around the world, particularly in Europe and the United States.

Despite all these problems, civil aviation continues to flourish and airlines are looking for record traffic levels for 1989. A total of nearly 1.2 billion passengers are forecast to travel by air on scheduled services during this year, an increase of around 7% on 1988 levels and representing a 60% increase over the decade from 1979. After the recession of the early 1980s the major carriers expect to make a combined net profit of 5,500 million US dollars, which again represents a steady increase on previous years.

A major factor in these increases is the effect of deregulation of

airline services in the United States. This has produced some turbulent effects in the industry but the final picture is now emerging of huge successful, conglomerate airlines. These include the Texas Air Corporation which owns Continental and Eastern, US Air which has taken over PSA and Piedmont, and others including American Airlines, United, Delta and Northwest. This pattern is of interest and concern to European airlines which are now faced with a similar scenario of deregulation in 1992. Already several of the major airlines are jockeying for position in acquisitions or takeover deals with other airlines. A typical example is SAS which has linked up with Continental Airlines in the United States and has bought a shareholding in Michael Bishop's Airlines of Britain Group which owns British Midland. Another Scandinavian airline, Maersk, has bought a shareholding in thriving Brymon Airways which, in turn, has taken over Birmingham Executive Airways. There is no doubt that the next few years will see major transformations in the shape and composition of the European airline industry — exciting times lie ahead.

Sharing in the air transport boom are the aircraft manufacturers who all have bulging order books. Included in this edition is a feature article on the European Airbus consortium which has gone from strength to strength in competing with the previously all-conquering American giants. Recent orders from major airlines such as Lufthansa and Cathay Pacific have assured the future of the new A330/340 family while the A320 is firmly set to become one of the best selling passenger jets of all time.

The most successful airliner to date is the Boeing 737 which has sold over 2,000 in various versions; the company took orders for a record 636 jet transports in 1988, substantially beating its previous record (set in 1978) of 461 orders. However, the company has not been without its problems. Delays in certification have set back the 747-400 programme although a flat-out effort enabled Northwest to take delivery of the first production aircraft in February 1989. The aftermath of the British Midland 737 accident dented customer confidence and called into question Boeing's ability to maintain quality while attempting to boost production to record levels — but the orders continue to pour in.

McDonnell Douglas also had a good year in 1988 with the MD-80 remaining a good seller and the new MD-11 attracting firm orders. At the end of 1988 total orders and sales of the MD-80 series had reached 1,353 aircraft, including options, and there were 88 firm orders for the MD-11. These figures have been boosted in 1989 by a massive order from American Airlines for up to 50 MD-11s and 100 MD-80s — one of the biggest airline orders ever placed. The expanding Air Europe has become the first British customer for the MD-11 with orders and options for up to 18 aircraft powered by Rolls-Royce RB211-524L engines. This latter order is significant because apart from showing the airline's continued expansion on to the long range market, it also shows the trend towards the use of larger aircraft in order to meet increased demand without adding to the airspace congestion problem. One MD-11 can carry the equivalent passenger load of three Boeing 737-200s.

Apart from the major airlines and manufacturers, this book also looks at the aviation scene in a wide range of situations. Geographically, the spread includes features on the Far East and South America as well as Europe and the United States. In scale the activities of aircraft range from helicopters and offshore patrol aircraft, through commuter airlines and manufacturers, to the big jet scene. All in all there should be something for everybody!

Leo Marriott

Tragedy in the Skies

The recent Pan Am 747 and British Midland 737-400 disasters will have considerable repercussions in terms of safety and security for the world of commercial aviation. ***Leo Marriott*** *examines the issues at stake behind two tragedies which between them claimed over 300 lives in the space of 18 days.*

AS COMMERCIAL aviation went about its business in the closing days of 1988, it could look back on the flight safety record of the previous 12 months with mixed feelings. On the one hand, a total of 54 fatal civil air transport accidents had killed 1,007 people — figures which, although regrettable, conformed with a steady (if unspectacular) increase in air safety when measured against the steadily increasing numbers of passengers. On the other hand there were a disturbing number of fatalities due to terrorist or military action and the worst of these occurred in July when an Iranian Airbus A300 (EP-IBU) was shot down by a missile from the USS *Vincennes* in the Arabian Gulf. A case of mistaken identity in an area racked by war caused the death of all 290 people on board the Airbus. However, the Gulf area seemed remote to most people in the United Kingdom who were spared the full horror of such a major disaster — for the moment.

At about 19.15hrs on the evening of 21 December 1988, a telephone rang in the office of the Duty Officer at the Department of Transport in London. The caller was the Watch Manager at the Scottish Air Traffic Control Centre (ScATCC) situated at Redbrae, not far from Prestwick Airport. The information which he passed on was brief and to the point: just after 19.00hrs local time, radio and radar contact had suddenly been lost with Pan American flight PA103 and reports from police and other sources indicated that it had crashed in the vicinity of the small Scottish town of Lockerbie and that wreckage was scattered over a large area.

The Duty Officer, no doubt shocked at this news, would have immediately contacted the Duty Co-ordinator of the Department's Air Accident Investigation Branch who in turn arranged for a team of investigators to be flown to the scene, the first stage of what was to be a lengthy and painstaking pro-

Above:
The broken nose section of Pan American Boeing 747 N739PA which crashed at Lockerbie on 21 December 1988 came to symbolise the extent of this enormous tragedy — Britain's worst air disaster. *Rob Taggart/Reuter*

cess to ascertain the cause of this terrible disaster.

The basic facts would have been immediately available from Air Traffic Control records which showed that the aircraft was a Boeing 747-121 (registration N739PA) operating a normal scheduled flight from Heathrow to New York. On board were a total of 259 passengers and crew who would have been relaxing as the aircraft cruised at 31,000ft and crossed the Scottish border at the Deans Cross VOR to pick up its Oceanic Track for the transatlantic crossing. Control of the aircraft had routinely passed to one of the sectors of the Scottish ATCC and would have shown up on their radar screens as a position symbol and alphanumeric data block showing the callsign, level and route details. Just after 19.00hrs, this data suddenly disappeared from the radar screens and repeated attempts by the alarmed controller failed to contact the aircraft.

On the ground, the full extent of the tragedy was slow to unfold. Although large pieces of wreckage were scattered over the Dumfriesshire countryside, a major part of the aircraft had come down on a residential part of Lockerbie, near the main A74 trunk road which runs from Carlisle to Glasgow. In the ensuing explosion and fire a number of houses were completely destroyed and others seriously damaged, while a massive crater was gouged into the earth. Fifteen of the townspeople lost their lives.

This was undoubtedly the worst ever air disaster in British airspace and consequently the accident investigation team was under intense pressure to determine the cause. It was obvious that something had caused the aircraft to break up in mid-air without warning and this pointed to only three possible events — a mid-air collision, a major structural failure or a terrorist bomb. The extent of the wreckage initially led credence to the first theory as it was thought that two aircraft must be involved, but it was quickly established that no other aircraft was in the vicinity at the time and that the pieces on the ground all came from the 747. Determining which of the other theories was correct was more difficult to establish and both held ominous consequences for future airline operations.

If the aircraft had indeed suffered a structural failure then the world's entire 747 fleet might have to be grounded and rectification of any structural faults would be a costly operation. The Pan Am 747 was in fact one of the oldest flying examples, the 15th off Boeing's production line, and had entered service in 1970. Several of the older 747s had already been withdrawn from service to have strengthening work carried out to skinning on the nose section — the very part of the aircraft where it was already established that the main failure had taken place. However, the accident aircraft had recently undergone a major overhaul which included extra strengthening of the airframe to allow it to act as a military transport in times of national emergency as part of the US Civil Air Reserve Fleet.

There is no doubt that, understandably, Boeing and the airlines should heave a momentary sigh of relief when it was announced that the initial investigation had proved that an explosive device, probably in the forward baggage hold, had been responsible for the break-up of the aircraft. This relief was, of course, short-lived as the implications of the security problems involved became apparent. Although PA103 had departed from London's Heathrow Airport, some passengers and their baggage had commenced their journey on a Pan Am feeder flight originating in Frankfurt, West Germany. At Heathrow their baggage would have been offloaded from the incoming Boeing 727 and transferred directly to the waiting New York-bound 747. Many of the passengers were US military personnel and their families travelling home for Christmas.

Although both airports prided themselves on their good record in security matters, it was obvious that the device (subsequently believed to be disguised as a tape recorder) must have been introduced into the baggage at one of the departure points. However the exact method, and the identity of the group responsible, have yet to be determined although it is almost certain that it was planted by a Middle East-based faction as a revenge for the shooting down of the Iranian Airbus earlier in the year. The fact that US intelligence sources were aware of a possible attack against a Pan Am aircraft was not widely disseminated and gave cause for complaint after the event.

Today, airport security is tighter than ever but there are always human failures. The Czechoslovakian Semtex explosive favoured by terrorist groups is extremely difficult to detect and can be moulded into virtually any shape — even formed as lining to an otherwise normal suitcase. The war against terrorism continues but it is not realistic to assume that the Pan Am tragedy will be the last of its kind.

Coming as it did only days after a major railway accident in Britain, the Lockerbie disaster had a greater emotional impact than might otherwise have been the case. But fate had another cruel blow in store as people were stunned to hear of another major accident on 8 January, less than three weeks after Lockerbie. This time the aircraft was a Boeing 737-400 operated by British Midland Airways which came down on to the M1, Britain's major trunk road, after experiencing engine problems on a flight from Heathrow to Belfast.

Although the collective reaction was 'Oh No! Not Again!', the two accidents had virtually nothing in common other than a tragic loss of life. The British Midland Boeing 737-400 was a brand-new aircraft which had just entered service. Approximately 15min after leaving Heathrow, and while climbing en route to Belfast, the captain reported to ATC that he had shut down the starboard engine due to vibration and a possible fire. Subsequently, after a radio discussion with the airline's operations headquarters at East Midlands Airport, the captain elected to divert to that airport so that the engine could be checked and the passengers transferred to another aircraft to complete their journey. As the 737 passed over East Midlands it was still at cruising altitude (25,000ft) and consequently the port engine power setting was reduced as the aircraft descended to the north before turning back to the field for positioning on to final approach for East Midlands runway 24. As the aircraft lined up with the runway at about eight miles from touchdown, the power was increased on the port engine which immediately exhibited signs of trouble with loss of power, fire warnings and pieces falling from the engine on to the ground

below. With power reduced to almost zero, the crew frantically struggled to maintain height and reach the runway, tantalisingly visible only a few miles ahead. But it was too late: the 737 hit the ground about half-a-mile short of the runway, bounced over the M1 motorway, and crashed into the embankment on the far side. The impact broke the fuselage at two points and fuel streamed out from the ruptured wing tanks. In an otherwise desperate situation there were two points of good fortune: nobody on the ground was killed or injured despite the fact that the motorway was busy at the time, and the East Midlands Airport Fire and Rescue services intelligently anticipated the accident site and were already on the move before the aircraft impacted. Consequently they were on the scene within two minutes and were able to smother the wreckage with foam in order to prevent the outbreak of serious fire while the first survivors were pulled from the shattered fuselage.

And, miraculously, there were survivors. Of the total of 126 people on board 47 were killed in the impact or died shortly afterwards, but the fortunate survivors were rescued and transported to waiting hospitals with commendable efficiency. Indeed it must be recorded that the work of the various rescue and emergency services at all the disasters around this unhappy Christmas and New Year period was beyond reproach and an inspiring example of what ordinary people are capable of when faced with a desperate need for their skills.

On the face of it, the AAIB investigators should have had an easy job in tracing the cause of the 737 crash. Both pilots had survived and the flight data and cockpit voice recorders were readily available. The wreckage was all in one place and easily accessible for inspection while there were any number of witnesses, both on the ground and among the survivors. Initial impressions were that the aircraft had suffered a failure in both engines and this raised serious questions concerning the safety of twin-engined aircraft. Based on experience with modern engines, most of the world's airworthiness authorities were clearing certain twin jets to fly long distances over water provided that they were within two hours' flying time of a suitable diversion airfield. If the British Midland aircraft had indeed suffered a double failure of its General Electric/SNECMA CFM56-3 engines then the whole concept of Extended Range Operations (Erops) would be called into question. As with the Lockerbie disaster, there was a lot of vested industry interest in the results of the investigation.

As soon as this began it was established that a double engine failure had not occurred and the starboard engine, which had been shut down early in the flight, was in fact perfectly serviceable. This fact was let slip by an AAIB investigator

Above:
The broken fuselage of the British Midland Boeing 737-400 lies on the embankment flanking the M1 motorway. Miraculously almost two-thirds of the 126 people on board survived the impact, including some from the battered tail section. *Leicester Mercury*

during a television interview and immediately fuelled all kinds of irrational speculation in the media — had the pilot incompetently ignored a fire warning on the port engine and shut down the wrong engine?

This hypothesis was patently absurd but, on the other hand, the reason for shutting down the starboard engine when it was functioning normally would obviously be central to the enquiry. One possibility was that the fire alarm systems had been incorrectly connected during assembly and consequently checks were ordered worldwide on all Boeing 737-400s and 300s. Subsequently, checks were also ordered on other Boeing types including the 757 and 767 and in several cases faults were found in fire suppression systems where cargo hold extinguishers were wrongly connected. However, no case of incorrectly fitted engine fire alarm systems was found, although one or two cases where the resultant extinguisher discharge would have acted on the wrong engine came to light, and by mid-February the CAA and FAA authorised the discontinuance of such checks.

Examination of the CFM56-3C1 engines removed to SNECMA's headquarters in France confirmed that the starboard engine had no fault at the time of the accident, although the port engine had suffered a broken fan blade prior to impact and this would have accounted for its failure on final approach. Following this the investigators renewed their examination of why the 'good' engine had been shut down initially. There had been reports of smoke on the flightdeck prior to the shutdown and it may be that the crew identified this as emanating from the starboard engine which supplied the majority of the cabin air. In the meantime the damage to the port engine caused by the broken blade may not have been enough to have immediately triggered the fire warning.

One factor looked at was the readout of the engine vibration monitor instruments, which were disliked by the airline's pilots. Reading on a scale of 0 to 5, they would often show a reading of 4 under normal operating conditions which meant that any deviation from this condition was difficult to spot. Thus a vibration increase in the port engine would not be immediately apparent, particularly if the crew had already formed the impression that the starboard engine was at fault and the port engine had been throttled back for the descent.

At the time of writing this article (February 1989) the results of the AAIB investigation had not been disclosed, so some of the above can only be considered as informed speculation. However it is certain that the accident will be shown to have been the result of a chain of happenings and circumstances rather than as a single catastrophic event as in the Lockerbie crash. The job of the investigators will be to establish the complete chain of events and determine which, if any, were the result of human error and which were purely mechanical. This is an extremely difficult process as the limits of each are not clearly defined. For instance, if the pilot misread or failed to extract information from the vibration meter then this could be termed 'pilot error'. However if the design and functioning of the instrument is shown to be less than ideal, is this a mechanical fault or is it human error on the part of the designers and engineers?

These questions will be considered and eventually a full report will clearly show what happened but, more importantly, will contain recommendations on how such a disaster can be avoided in the future. This will take time, although a preliminary report will identify the causes. In the case of the British Airtours Boeing 737 which caught fire on the runway at Manchester in 1986, the final report and recommendations were not published until 13 March 1989.

In the meantime it is inevitable that there will be increased anxiety about the safety record of commercial aviation, and in particular Boeing has suffered from a severe loss of prestige. Despite this, airlines are looking forward to a continuing boom in air transport and all the major manufacturers are hard-pressed to meet the demand for new aircraft. These pressures are seen as having a detrimental effect on air safety, and the performance of airlines and manufacturers will be closely monitored in the years to come. Statistically air travel is safe but, by their very nature, aircraft accidents are spectacular and attention-grabbing which leads to a perception of increased risk. Those who work in the aviation industry will realise the tremendous effort which goes into maintaining safety as a first priority in all circumstances, but they will have to labour hard to preserve that image with their passengers and the world at large.

Below:
The 737-400 was a brand-new variant of the best-selling Boeing 737 and the first example only entered service with Piedmont Airlines in September 1988. Compared to the previous 300 model, the new variant has a fuselage extended by 104in to give a maximum capacity of 172 seats; maximum weight is up by 4,000lb and aerodynamic improvements increase cruising speeds by up to 18kt. *Boeing*

Flying Dynamite

TNT 146 Freight and Parcels Operations

Thomas Nationwide Transport — commonly known as TNT International Aviation Services — and British Aerospace have pooled their expertise to achieve a unique operation for the transportation of freight and small packages round the clock. ***Philip J. Birtles*** *looks at TNT's operations.*

WHEN TNT made the decision to expand its road transport organisation throughout Europe to include air transport, the BAe 146 QT was a key part of the programme, due to its low noise levels allowing unrestricted night operations, and excellent economy. Instead of opting for cheap first-cost secondhand piston-engined air freighters with their high maintenance costs, TNT took the plunge and placed a commitment for 72 of the specially adapted BAe 146 QT (Quiet Traders) based on the standard series 200 airframe.

The initial order was placed on 1 December 1986, the first aircraft going into operation on 5 May 1987. By mid-1988 five aircraft were in service — two in the UK operated by Air Foyle, one in France temporarily on the British register as G-TNTH, one in Sweden (SE-DEI) and another in Italy (I-TNTC). In each case, the aircraft are operated by an established air operator in the country concerned. This fleet flies regular services from Britain, Spain, France, Germany, Italy, Hungary, Austria and Scandinavia to the hub at Nuremberg in Germany. Five more 146s were added to the

Above left:
The BAe 146-200 has been adapted to the Quiet Trader (QT) configuration as a dedicated freight aircraft. For this purpose a large freight door has been fitted behind the wing trailing edge and most of the cabin windows have been deleted. *BAe*

Above:
TNT was the launch customer for the BAe 146 QT with a total commitment of 72 aircraft, taking all the production of this variant over a five-year period.

European fleet by early 1989, with an expected final growth to some 20 aircraft. The remaining aircraft in the commitment, tying up the 146 QT production lines for five years, will be absorbed by TNT's interests worldwide, with Ansett in Australia (already a 146 operator in the normal passenger configuration), North America, South America — particularly Brazil — and throughout Asia.

Some statistics will help to illustrate the scale of the TNT operation: the company operates 200 depots in 16 West European countries while some 5,000 vehicles are in use with 9,000 employees providing the collection, consolidation, loading and delivery services.

The object of the TNT system is to move freight at any time and the 146 QTs are used to support the European overnight door-to-door service. A call to any one of the depots for a late afternoon collection allows for a door-to-door delivery service before 09.00hrs the next day. There is a money-back guarantee in case of late delivery and TNT has achieved a better than 95% reliability within the first year of operation with the BAe 146, still including the road transport link.

The overnight air express service covers packages between one and 50kg, while the Elite service covers heavier loads up to 500kg. A maximum single load on the aircraft is limited to 4,000kg.

To make a full utilisation of the aircraft during the daytime, the air container line was started in September 1988 providing a scheduled freight service hubbing through Cologne and covering most major European centres. An innovative ingredient in the daytime operations is the sale of capacity by container load, there being room for six-and-a-half containers on each aircraft, each full container being capable of carrying up to 1,600kg or 9.25cu m. This door-to-door operation with one company was only feasible because TNT had already developed the surface transport network.

With the overnight service, the aircraft finish up at the end of each spoke giving additional opportunities for *ad hoc* charters, including the rapid transportation of race-horses around Europe. The user is able to choose from a range of services for his packages, from the overnight/12hr service known as the Elite, to Air Containers Line which could take 24, 36 or 48hr depending upon need, and the 72hr road economy service.

The airfreight centre for TNT in Britain is Luton Airport, where the company moved from Birmingham in mid-1988. Domestic air services are flown into Luton from Prestwick, the main maintenance base, and Belfast. There are 27 road freight depots, including Luton, throughout the UK with Atherstone in the Midlands being the main centre where all loads are delivered for road distribution. The major administrative base is at Northampton where a mainframe computer is located and used for the compilation of the cargo manifests.

Each BAe 146 load consists of up to six-and-a-half 'igloos', each prepacked and routed to a specific final destination airport. The average turnround time is 30min, but a fastest time of 9min has been recorded.

The BAe 146 aircraft are achieving around 2,800 flying hours/year, which will increase as the daily utilisation is improved upon by the Charter Division of TNT. The major requirement of the day charter work is that the aircraft is available on time for its night-time schedules.

TNT 146 QTs are fitted with a large upward-opening freight door on the port side of the fuselage behind the wing. Most of the cabin windows are blanked off and the cabin is an unfurnished plain-trimmed cargo area. Additional space for loose cargo is in the normal underfloor holds. The crew entry vestibule

in front of the main freight hold is a somewhat functional area with a folding seat, coffee machine and toilet.

The two-crew flightdeck is fitted with a single Category 2 autopilot giving a decision height of 100ft and a Runway Visual Range (RVR) of 350m. The Universal Navigation System (UNS) works on the Distance Measuring Equipment (DME) giving 40,000 way-points. Omega navigation system is also fitted.

As mentioned earlier, the initial pair of British-registered aircraft, G-TNTA and G-TNTB, are operated by Air Foyle which has 14 crews available, in addition to the Chief Pilot. The aircraft normally cruise at Mach 0.65 up to the maximum cruise ceiling of 31,000ft, and limit the airspeed to 295kt.

On a typical sortie, the 146 will fly from Luton to Nuremberg in 1hr 20min, taking off 37min later for Cologne. After a longer stop at Cologne for an hour, due to the late

Below left:
The overnight packages are packed in igloos allocated to specific destination airfields, before loading aboard the BAe 146. *BAe*

Below:
Protection to the loading team in bad weather is afforded by the upward-opening freight door. The igloos are manhandled from the towed trolleys on to the mechanical handling equipment, which raises them up and into the aircraft. *BAe*

Bottom:
The cabin of the BAe 146 QT is fitted with roller mats for ease of cargo movement, and can be used to carry small packages in igloos, or larger loads on pallets. *BAe*

arrival of one container, the 146 then returns to Luton after one hour airborne.

The BAe 146 has been found to be ideal for the task. It has achieved a high level of reliability and is well liked by the air and groundcrews. Loading is rapid and straightforward, as is also unloading. The aircraft, due to its quiet operation, has no night noise restrictions and it qualifies for cheap landing rates. Nuremberg is an ideal European hub due to its excellent facilities and lack of congestion from other overnight traffic.

TNT is therefore poised to take full advantage of the major growth market in overnight and *ad hoc* freight throughout Europe in preparation for the relaxation in customs regulations in 1992. This relaxation will improve freight transition and business as a whole, increasing the demand for air transport of freight and, therefore, the demand for more aircraft capable of being acceptable to the community during the hours of darkness.

The BAe 146 and TNT air freight partnership is expected to continue to expand in response to the anticipated rapid growth in demand.

Above:
Nuremberg makes an ideal European hub, where it is not uncommon to see up to four aircraft loading and unloading at the same time. *BAe*

Below:
Daytime freight operations include the rapid transportation of racehorses from one meet to another throughout Europe. The horses arrive in the best possible condition to race, and the cost of the transport is often more than paid for by the resulting prize money.

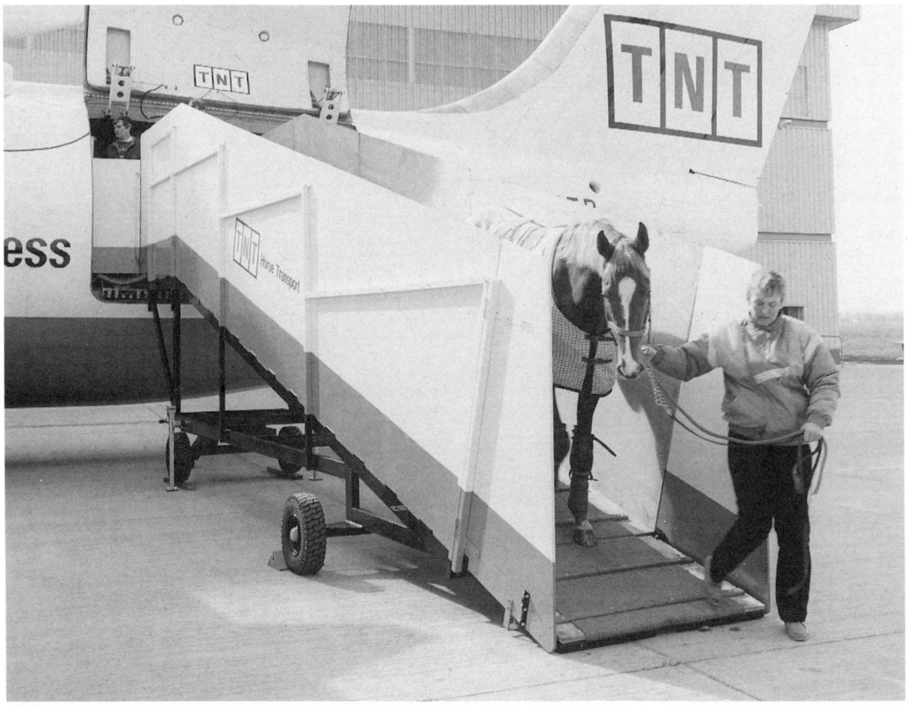

Watchdogs Over the Sea

***Chris Stock** takes a look at offshore surveillance and monitoring by air which is proving to be vital to the well being of the environment and economy.*

THE MARITIME nations of the world have relied upon the sea as a substantial source of revenue whether it be from the fishing industry, tourism or latterly, offshore oil extraction. Over the years, there has developed the need to protect these resources against the predatory behaviour of other nations and the indiscriminate dumping of waste materials into the water. Dwindling fishing stocks have brought stricter regulations and controls over the fishing fleets with the consequent need to monitor their activities.

Oil pollution remains a major hazard for coastlines and wildlife which needs to be dispersed at the earliest possible time to prevent lasting damage to the environment.

The extension of the fishing limits to 200 miles created large areas of sea under national control which require surveillance and monitoring to ensure the national interest is protected. Most countries achieve this function through the combined use of surface vessels and aircraft. For a number of years, the role of maritime surveillance fell within the compass of national defence forces. However, for reasons of cost and higher priority tasks, surveillance of the fishing fleets and possible sources of oil pollution has been transferred in increasing numbers to civil contracts by the respective government departments responsible. Britain is no exception to this trend. When the fishing limit was pushed out to 200 miles in January 1977, the total area of sea that required patrolling exceeded 250,000sq miles. In the early stages, the task of monitoring and surveillance was delegated to the RAF which used Nimrods of No 18 Group. Under the aegis of the Ministry of Agriculture, Fisheries and Food (MAFF), the Nimrods were tasked for 180hr/month fishery protection duties. However, full coverage of the UK waters was not possible and consequently Sea Devons of 781 Squadron, FAA, provided surveillance in the English Channel, South West approaches and part of the Irish Sea on an *ad hoc* basis.

The FAA squadron disbanded in March 1981, leaving part of the UK

Below:
The Department of Agriculture and Fisheries for Scotland (DAFS) now employ a specially-equipped Fokker F27 for long endurance patrols. The aircraft is based at Edinburgh and operated by Harvest Air *Scottish Information Office*

area of responsibility without aerial surveillance. At this juncture, MAFF considered various options, including the use of civil aircraft, and it was decided to experiment with a civil contract which was put out to tender to cover the unprotected waters. A company based in Dorset, Flight Refuelling Ltd (later to become FR Aviation) was awarded the contract to provide maritime surveillance with a Pilatus Britten Norman Turbine Islander. The contract proved to be extremely successful and inspired MAFF to reconsider its airborne surveillance policy. As the surveillance task evolved with experience, FR's Islander assumed the role of inshore (out to 50 miles) surveillance extending around the UK coastline.

The economics of using cheaper civil aircraft was not lost on a money-conscious MAFF, particularly when comparing the high hourly costs of the sophisticated Nimrod with its 13-man crew. Although the inshore area was now covered, the crucial remaining area from 50 miles out to 200 miles required a cost-effective aircraft capable of long patrols, carrying radar and navigation equipment appropriate to the surveillance and monitoring task.

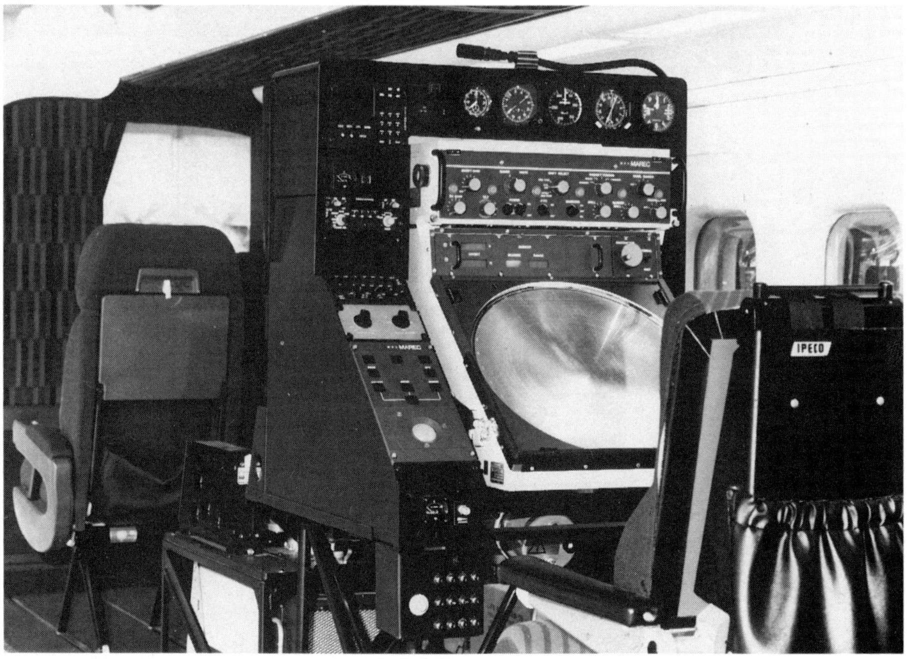

Top:
This appropriately registered Cessna 404 Titan is also based at Edinburgh and flown on behalf of the DAFS. *Scottish Information Office*

Above:
MEL's lightweight MAREC radar is typical of the equipment available for use by civil aircraft employed in offshore patrol duties. This neat installation is aboard a Dornier 228 for the Indian Coastguard service.

Right:
G-MAFF is a Pilatus Britten Norman Turbine Islander operated until recently by Flight Refuelling at Bournemouth on behalf of the Ministry of Agriculture and Fisheries. It has now been replaced by the faster Dornier 228s. *Leo Marriott*

When it came to the choice of aircraft, MAFF was faced with an embarrassment of riches. Every aircraft manufacturer had anticipated the maritime surveillance potential and was offering model adaptations from their product range. A quick check reveals nearly 40 different types available, ranging from popular and cheap twin turboprop/piston right up to large turboprop aircraft. MAFF evaluated the options and decided on the Dornier Do228-200; four were ordered. One was destined for the Falkland Islands (operated by Bristows); the other three were prepared for UK maritime surveillance, one was leased for operation in Scottish waters whilst the other two covered the remaining area.

The UK aircraft were equipped to MAFF specification which included a Bendix RDR1500 360° scan radar, and Racal R/NAV 2 with inputs from Decca, Doppler, Omega, and VOR/DME. Information is presented to the crew on a tactical display situation between the two pilots. The Scottish aircraft was operated by Harvest Air whilst FR Aviation were contracted to fly the two remaining Dorniers. Subsequently the Scottish Dornier was replaced by the unique Fokker F27 which was extensively re-engineered for the surveillance task by Air UK in late 1987. The military Ferranti Seaspray Mk 3 search radar was installed along with a Litton LTN-72R Inertial Navigation system. This aircraft is based at Edinburgh Airport flying daily offshore surveillance missions.

Whilst monitoring the fishing fleets is vital to the UK economy, pollution of the seas remains an environmental hazard with major consequences for the coastlines, wildlife and the food chains in the ocean. One of the most serious pollutants is oil and the main sources of pollution are from ships and offshore installations. In particular, the North Sea has become the focus of the environmental lobby and the governments of the North Sea nations are taking combined action to keep the damage from pollution within acceptable limits.

On behalf of the UK Government, the Department of Transport has formed a Maritime Pollution Control Unit (MPCU) which is for the identification and dispersal of oil pollution. As with aerial surveillance for the fishery protection, the MPCU has decided that aircraft are the most cost-effective way of monitoring oil pollution. Consequently, the MPCU sponsored a one-year trial which started at the beginning of the 1988 financial year using an Air Atlantique Cessna 402. This aircraft is fitted with sideways-looking radar which gives coverage out to 30 miles either side of track. Using this radar, the crew can identify oil slicks within radar cover and estimate the shape and size of them. The radar coverage enables the aircraft to monitor large areas of the ocean within a given flight pattern.

Although the Department of Transport is responsible for this trial, it shares the aircraft hours with the Department of Energy which has responsibility for all offshore installations within the UK sector. The total monthly hours flown are 50 with the two departments sharing them — Energy taking 30hr and Transport 20hr. The UK works closely with its European partners, particularly Holland, Norway and Germany,

monitoring the North Sea for oil pollution and thereby giving complete coverage of an environmentally sensitive area of water.

Although Cessna 402 trial was scheduled for 12 months, the Department of Transport envisages that aerial surveillance will continue on a permanent basis. Careful evaluation of the trial may indicate that changes might need to be made. However, as with the fishery protection aircraft, there is no shortage of aircraft to undertake the role. What is certain is that the contract will be given to a civil operator.

In the UK, maritime surveillance had been directed at two main areas — fishery protection and oil pollution — but it may eventually be extended to cover other antisocial activities such as drug smuggling and the illegal importation of immigrants. Other countries have been successful in enforcing the law by the use of aircraft and it may well be only a matter of time before the UK follows suit.

The use of civil aircraft in the offshore tasks is increasing; the benefits to the government are reduced costs and the establishment of dedicated operators who, through continuing experience, provide an extremely efficient service. The advantage to the RAF is that the civil operator has released the maritime force from an onerous task and is allowing the Nimrods to perform their primary defence tasks. However, aircraft are unable to carry out the complete operation which includes the boarding of vessels, and consequently a complementary surface vessel fleet is essential, although its size can be small through the reduced equipment of surveillance. In this context it is interesting to note that the French have carried out trials using a British airship carrying an inflatable boat which can be lowered to the surface for boarding operations.

Offshore surveillance and monitoring by air has shown to be vital to the environment and economy. The advent of the civil aircraft and modern radar equipment into the operation has been a successful and cost-efficient exercise and the watchdogs have become indispensable in protecting the nation's interest and environment.

Below:
The Skyship manoeuvres to drop its boat to facilitate an interception. Note the shrouded propellers. *Airship Industries*

Above:
In UK waters the various government-sponsored patrol aircraft work in close co-operation with ships of the Royal Navy's Fishery Protection Squadron. HMS *Guernsey* is one of seven 'Isles' class Offshore Patrol Vessels deployed on such duties. *Leo Marriott*

Right:
The increasing economic importance of offshore activities has led to a requirement for effective patrolling of these areas. Civil aircraft, such as this Turbine Islander, are playing an increasingly important part in this role.
Pilatus Britten Norman

Below:
One drawback of using aircraft for patrols is that it is not possible to stop and board a suspect vessel. To overcome this problem Airship Industries modified its Skyship 600 to carry an inflatable boat and trials were carried out by the French Authorities. *Airship Industries*

Hong Kong Choppers

*South East Asia is one of the fastest growing areas of the world in economic and industrial terms, and playing a vital part in this growth is civil aviation. Globetrotting photographer **Robbie Shaw** investigates the activities of Heliservices (Hong Kong) Ltd, the Colony's only commercial helicopter operator.*

ALTHOUGH the Territory of Hong Kong is only 1,067km sq, the helicopter is a tremendous and vital asset. Travelling around Hong Kong can be a very tedius experience, ranging from the continuous traffic jams in the heavily populated commercial areas of Hong Kong Island and Kowloon, to the natural obstacles of numerous hills, mountains and the 235 islands of the Territory. Consequently the helicopter has become a part of life in Hong Kong. The RAF and Army Air Corps both maintain a squadron of helicopters in the Territory in support of ground troops, whilst the Royal Hong Kong Auxiliary Air Force's (RHKAAF) SA365 Dauphins are extensively used, including a weekly 'flying doctor' service to the islands and remote parts of the Territory. There are also three Hughes 500s owned by local businessmen, one of whom uses his machine as a mode of transport to and from work every day.

The only commercial helicopter operator in the Territory is Heliservices (HK) Ltd, which has a maintenance facility at Sek Kong airfield in the New Territories close to the border with China. Sek Kong is a military airfield, home to the Wessex of No 28 Squadron RAF and No 660 Squadron AAC. The location of Sek Kong is a good example of just how useful the helicopter is in the Territory. Flying time from the airfield to Fenwick Pier Street heliport in Central Hong Kong is a brief seven-minute hop, yet this journey can take 90min by road!

The Fenwick Pier Street heliport is operated by Heliservices and is centrally situated on the waterfont of the commercial area of Central District. Also located in Central District is the company's Head Office, administered by General Manager John Simler, while Chief Pilot Capt Doug Cavannagh oversees the Sek Kong operation. The Chief Engineer, Graham Marsh, is assisted by a licensed engineer and a mechanic.

Capt Cavannagh is no stranger to Hong Kong as he was a pilot with the company a few years ago when it had a fleet of five aircraft; prior to that he piloted Wessex HC2 with the RAF's No 28 Squadron.

Heliservices was initially formed as a joint venture between Evergreen Helicopters in the USA and a local Hong Kong family in order to carry out a power line construction project for one of the Territory's utility companies. This project lasted approximately six years, during which time over 16,000hr were flown using four SA315 Lama helicopters. The helicopter was found to be necessary for this type of construction work in Hong Kong because of the problems and cost of constructing roads for vehicular access in the rugged terrain. In addition, some of the work had to be carried out in National Parks and road construction, of course, would have ruined great tracts of these. The Lama was found to be ideally suited to this type of work, requiring as it does small landing areas; it also has a high payload and is efficient and economic to operate. At around this time the company also operated an Alouette 3, Ecureuil and Jet Ranger.

Today the fleet comprises one SA315 Lama (VR-HIP c/n 2569) which the company has had on its books since 1979, and one AS350B Ecureuil (VR-HJD c/n 1495) which was purchased early in 1986 to replace the Bell Jet Ranger which was held on lease. Business is currently transferred from contract to *ad hoc* charter. The Lama is largely used in association with construction work, moving equipment to remote and inaccessible hilltops — a task for which it is ideal due to its ability to lift 1,000kg payload. Important customers for the aircraft in this role are China Light & Power and the Hong Kong Telephone. It is frequently cheaper for a construction company to hire the Lama for lifting than it would be to hire a crane or even manual labour on occasions — a fact that cost-conscious local managers are slowly becoming aware of. At the time of writing the Lama is being utilised by a company which is carrying out core sampling prior to construction of a new Lion Rock tunnel. Towards the end of 1987 Heliservices is hoping to land contracts for more construction-related work, including constructing sites for a new Harbour Radar system and a nuclear power station. If these come to fruition the company will be looking for two more Lamas.

The Ecureuil is the busier of the two machines. It is cheaper to operate than the Lama and is used extensively for aerial photography and by movie film crews. The latter seem to be in abundance in Hong Kong! Heliservices also offers the tourist the unique opportunity of viewing and photographing the entire Territory in style and comfort; scenic tours are flown from the heliport at Fenwick Pier Street. The duration of these tours ranges from a short 15min around-the-harbour trip, to a full one-hour flight over the harbour, outlying islands, and a jaunt along part of the Chinese border. Hong Kong from the air is a sight to remember as the contrast between the crowded harbour, 'concrete jungles' and the green hills and duck farms near the border is stunning. An increasing number of tourists are taking advantage of these trips — although it has to be said that the majority of tourists come to Hong Kong with only one thing in mind: shopping!

Only a small percentage of tourists to the Territory are interested in seeing the sights and culture. An effort is therefore being made to attract even more airborne sightseers with the establishment of international agency link-ups. A service recently introduced is the 'Heli-Golf tour' in conjunction with the Clearwater Bay Golf & Country Club, which is aimed at visiting businessmen. By using the helicopter, a full round of golf can be enjoyed followed by lunch and still leave plenty of time for those business meetings. A daily weekday service is provided to the Club which is one of the most spectacular courses in Asia. Another recent introduction is 'Heli-Water tours' on Lantau Island, the largest island in the Territory, which is even bigger than Hong Kong Island itself. Here, after a scenic flight through the harbour, tourists can be left to enjoy water-skiing and other aquatic sports.

Major factors which decided the purchase of the Ecureuil were Heliservices' wish to standardise on Aérospatiale equipment, the belief that it is the best single-engined four/five-seat helicopter available, and its competitive price. Monthly utilisation of the fleet currently runs in the region of 80-100hr, the Ecureuil accounting for about 60% of this; the Lama accounts for the remaining 40%. The company is currently looking into the possibility of operating a twin-engined machine which would enable it to extend services to public transport, night and overwater operations.

There is no doubt that Heliservices is a progressive company which can provide a much-needed form of public transport in Hong Kong and, if the hoped-for contracts in the construction industry materialise, then the company is assured of a rosy future.

Below:
Powered by a Turbomeca Artouste IIIB turboshaft, the Aérospatiale Lama is based on the well-known Alouette design and can carry a pilot and four passengers, or slung loads of up to 2,500lb.

Bottom:
The Aérospatiale AS350B Ecureuil is faster and more sophisticated than the Lama. It is particularly popular with film and TV companies for aerial photography over the Crown Colony.

écureuil
AS 350 B
DANGER
VR-HJD

Above:
Heliservices Ltd is the only commercial helicopter operator in Hong Kong and two of its aircraft are shown over the company's base airfield, Sek Kong.

Below:
Tourist and business flights are made from the downtown Fenwick Pier Street heliport using the five-seat Ecureuil.

Flying Down to Rio

Rio is renowned for its vibrant carnival and its celebrated beachside areas of Copacabana and Ipannema, immortalised on celluloid and in the lyrics of popular songwriters down the years. However, it also boasts two sizeable civil airports. ***Leo Marriott*** *takes up the story where Fred Astaire and Ginger Rogers left off . . .*

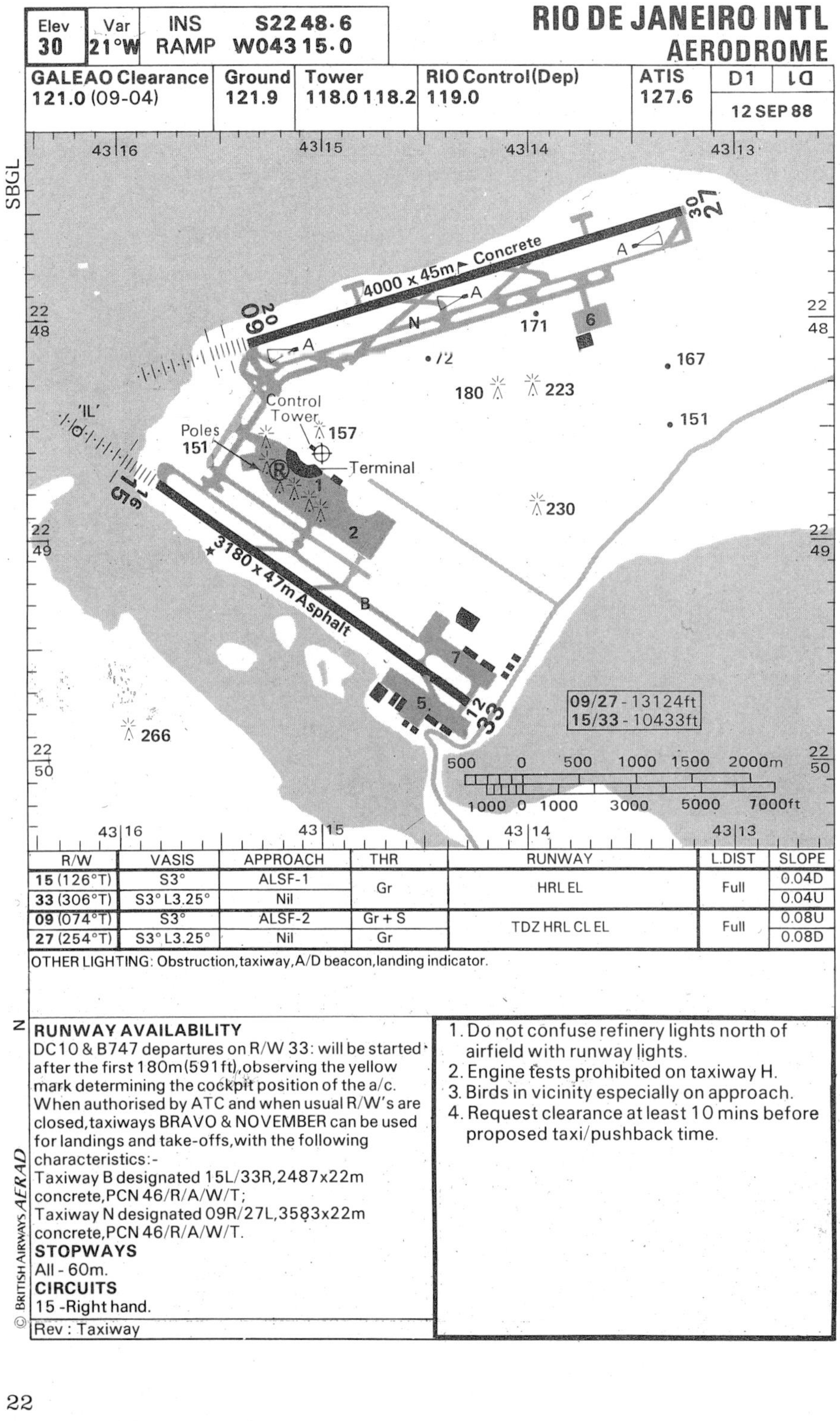

RIO DE JANEIRO is one of the world's great cosmopolitan cities with a population of over 10 million. As well as the renowned beachside areas of Copacabana and Ipannema, there are also the bustling downtown business sections, industrial zones and the largest port in South America. Although the newly-built city of Brazilia became the nation's official capital in 1960, Rio is still of vital importance and is only exceeded in size by the great conurbation of São Paulo which has a population of over 15 million and lies some 300 miles to the southwest, along the coast.

Rio's main civil airport is the Aeroporto do Galeao which is constructed on partly reclaimed land surrounding a small island in the Baia de Guanabara just north of the city. This was officially opened in 1977 and is reached by a short causeway from the mainland. It features two main runways (RW09/27 — 13,124ft long, RW15/33 — 10,433ft) laid out in an open 'V' plan with the area in-between occupied by a modern terminal building and a vast parking apron. At the western end of the airport is a separate apron complex used by the Brazilian Air Force. The terminal building is in the shape of a large semi-circle with service facilities at ground level, arrivals at first floor level with check-in desks, and departure lounges on the level above. The third floor level houses offices, restaurants and a panoramic spectator's gallery running the full length of the building. Work is commencing this year on the construction of a second terminal which will cost $250 million and handle 7.5 million passengers a year.

From the United Kingdom, Rio is served by direct flights from Heathrow operated by British Airways and VARIG, the major Brazilian

Left:
Rio de Janeiro International Airport. *British Airways AERAD*

Above:
Rio is a destination for most South American airlines and a rich variety of operators can be seen at various times. A typical example is this Boeing 707-320B (ZP-CCE) of Lineas Aereas Paraguayas — Paraguay's state airline which also operates three DC8 Series 60s, two more 707s and an Electra.
All photographs by Leo Marriott

international airline. British Airways flight BA245 operates three times a week, arriving in Rio at 06.10 local time after an 11hr flight from Heathrow. To cater for a long sector such as this the crew is reinforced by an extra pilot and two bunks are available in a rest area immediately behind the flightdeck. In the after section of the passenger cabin an enclosed rest unit is provided for the cabin staff, displacing some of the centre rows of seats in economy class. This latter unit is inevitably known as the 'Wendy House'.

After a short stay on the ground at Rio, the aircraft continues on to its final destination at São Paulo, returning in the afternoon to operate the overnight service (BA244) to London. As the original crew will have used up their statutory duty hours on arrival at Rio, they rest over for two nights and then operate the Rio-São Paulo-Rio return flight on the third day. A further two-night stop-over follows before the crew finally return to London with the BA244. These enforced stop-overs, although necessary from the point of view of flight safety, are nevertheless much enjoyed by the crews rostered to this route.

VARIG (Viacao Aerea Rio-Grandense) is the other scheduled carrier on the Rio-London route and has a

Below:
One of VARIG's Electras on the apron at Rio's Santos Dumont airport. The continuing growth in demand for seats on the popular shuttle service to Sao Paulo means that the Electras will shortly be phased out of service in favour of jet flights from the larger, but less convenient, Galeao airport.

history stretching back to 1927 when it was formed to operate services within Brazil using Dornier flying boats. In the period up to 1945 the airline expanded only gradually, using Junkers Ju52s, de Havilland Rapides and Lockheed Electras. After World War 2, DC-3s and C-46 Commandos were obtained in substantial numbers enabling a network of international services to be built up within South America. However, VARIG really moved into the major league when it commenced operations from Rio to New York using Lockheed L-1049G Super Constellations in 1955. The first jet equipment was two Caravelles which were delivered in 1959 and, amazingly, were used on a complex route of seven sectors from Buenos Aires, through Rio, to New York. However, a year later the airline took delivery of its first pair of Boeing 707-441 long-range jets which were more suited to this intercontinental route.

A further major expansion took place in 1961 when VARIG was asked by the government to take over one of its rivals, REAL, which operated a large mixed fleet including Electras, Constellations, CV440s and DC-6s. This instantly made VARIG the largest airline in South America, a position confirmed in 1965 when Panair do Brasil was also absorbed. The late 1960s and early 1970s were a period of expansion and modernisation. With the delivery of the first of 10 HS748s in 1967, the airline started to phase out its remaining piston-engined equipment while new jet equipment — including Boeing 727s, Boeing

Above left:
British Airways Boeing 747-236B *City of Manchester* operating Flight BA244 arrives at Rio's Galeao airport from Sao Paulo ready to operate the return flight to London later in the evening. *All photographs by Leo Marriott*

Left:
Transbrasil is a privately owned airline with a fleet of 26 Boeing jets of various types including the 737-300 and 727-100 shown here. The colourful tail markings are an eye-catching feature of these aircraft.

Top:
One of VARIG's two Airbus A300-B4s positions on to its parking stand. In the background on the left can be seen part of the city of Rio and the cranes of its massive port, with the famous Sugar Loaf Mountain rising beyond.

Above:
A beautifully preserved AT-6 Texan, one of several ex-Air Force examples in private hands.

737s and DC-10-30 aircraft — was delivered from 1970 onwards.

In 1975 another airline, Cruzeiro, was taken over by VARIG, although allowed to retain a separate identity. In the past decade VARIG has followed a worldwide trend towards larger high-technology aircraft with orders for Airbus A300s, Boeing 747 and 767 aircraft. Latest orders include the MD-11 and Boeing 737-300.

Cruzeiro also operates from Rio's Galeao airport and has a fleet consisting of two Airbus A300B4, six Boeing 727-100 and six Boeing 737-200. It operates a route network within Brazil and to Bolivia, Surinam, French Guiana, Uruguay and Argentina. Like VARIG, its origins can be traced back to the 1920s when it commenced operations with Dornier flying boats and later with Ju52s and Focke-Wulfe Fw200 Condors.

Two other major Brazilian airlines account for a significant proportion of Galeao's air transport movements — VASP and Transbrasil. The latter operates a mixed fleet of Boeing aircraft (707, 727, 737 and 767) in a striking all-white livery with a dazzling multicoloured striped pattern on the tail. Privately owned by

Above:
VARIG operates a total of 16 Boeing 737s including four of the 300 version, one of which is shown on the apron at Galeao airport.

Below right:
One of Cruzeiro's two Airbus A300B4-200 aircraft starts engines after pushing back from its parking stand. The airline is wholly owned by VARIG and also has six Boeing 737s and six Boeing 727s on strength.

the Fontana family, the airline is basically a domestic operator (apart from some international cargo flights) but is looking for international rights to expand its business.

VASP (Viacao Aerea São Paulo SA) is also a domestic airline and is owned by the São Paulo regional government. Its fleet includes no less than 28 Boeing 737 aircraft of various versions, as well as three Airbus A300B2 and two 727s. It has extensive investments in hotels, car hire, travel agencies and other travel-related industries and offers an extremely efficient service at all levels throughout Brazil.

Although Galeao is now Rio's major airport, the older Santos Dumont downtown airport is perhaps better known. Originally developed in the 1930s on reclaimed land only a few hundred yards from the commercial centre of the city, it is now far too small for modern jet operations. However, it is still a bustling terminus for the famous Rio-São Paulo airbridge flown by the ageing fleet of Lockheed Electras wearing VARIG's colours. In fact the Ponte Aerea (Air Bridge) is run by a partnership of VARIG, Cruzeiro, VASP and Transbrasil with each airline allocated a fixed percentage of seats as a result of a government-sponsored agreement (VARIG 52%, Cruzeiro 19%, VASP 22%, Transbrasil 7%). A total of 14 Electras are available and at peak times a flight departs every 20min. Normal cruising altitude on this route, in both directions, is 16,000ft (Flight Level 160) and this necessitates a parallel airways system between the two points. Flight time is 50min to Congonhas airport at São Paulo, also situated in a downtown area and surrounded by high-rise buildings.

The unique situation and size of both airports limits the potential for jet equipment on this route and trials have been carried out using Boeing 737s and the BAe 146, while the Fokker F100 has also been considered. However, none of these aircraft completely fits the bill although a substantial increase in capacity is required to meet the ever-growing demand for seats (currently almost two million a year). Consequently it has been decided that the service will be operated from Galeao airport using VASP Boeing 737s and, unless there is a late change of plan, 1989 will be the last year of operation for the Electras from Santos Dumont.

This does not mean that the airport will close. On the contrary, it will continue to provide a useful service to the business community as its already busy general aviation activity increases. In addition, several Brazilian regional operators will continue to fly from here. These include Rio Sul which is a VARIG subsidiary and operates a fleet of F27s and Bandeirantes, currently being replaced by new Embraer Brasilias, on commuter routes to the south of the country. Another operator is TAM (Transportes Aeros Regionais) which also operates F27s and Bandeirantes and is a subsidiary of VASP which holds two-thirds of the company. Finally there are TABA, which covers the northwest of Brazil, and VOTEC. Both companies also use F27s and Bandeirantes, although the latter also operates a large helicopter fleet in the offshore oil support role and may abandon its domestic services.

Santos Dumont airport is, of course, named after the famous Brazilian aviation pioneer who was responsible for much of the impetus

to early powered flight in Europe; he built and flew a variety of machines in France during the first decade of this century. Apart from the name of the airport, he is also remembered at the small hill town of Petropolus, some 30 miles north of Rio, where his house is preserved and maintained as a small museum. The airport itself is overlooked by the famous Sugar Loaf Mountain, one of Rio's great landmarks. However, this presents a major obstruction to the departure and arrival paths of aircraft using the airport, as anybody who saw the James Bond film *Moonraker* will remember. It is this, coupled with the relatively short runway (4,399ft) and lack of parking space which has precluded the effective use of jet traffic.

Apart from the commercial airports, Rio also has a few general aviation fields, although considerably less than a comparably sized city such as London. One of these, Jacarepagua, is near the new town of Barra situated beside the coast to the southwest of Rio. Built on the flat coastal plain, it has a single tarmac runway and is home to almost 100 aircraft. Several of these belong to the Aeroclube do Brasil, a national organisation which also provides a reserve commitment to the Brazilian Air Force. It operates a variety of aircraft including versions of the Fokker S11 and Piper Warrior, both produced at various times under licence in Brazil.

There is an FBO providing engineering facilities for a variety of aircraft including Bandeirantes of the Air Force, and helicopters. At the other end of the airfield is a range of new hangars and workshops which provide homes for private and business aircraft, an aircraft sales company and a couple of factories turning out microlight aircraft of various designs. Flying the latter is a popular activity in Brazil and there is a thriving microlight training school just to the south of the airfield which operates almost daily from dawn to dusk. Hang-gliding is also very popular, with pilots launching themselves off the 3,000ft peaks behind the coastal resort of nearby San Conrado and remaining airborne for hours using the constant sea breezes which generate substan-

tial upcurrents as they blow over the mountains. San Conrado in fact boasts Rio's smallest airport — a grass area less than 100sq yd, equipped with a windsock and situated just above the beach — used by the hang-gliders as the lift finally runs out and they come down to sea level.

Altogether Rio lives up to expectations, not only as a business and tourist centre, but as a thriving centre for civil aviation — a force which has done much to assist Brazil's emergence as one of the world's major trading nations.

Above:
Lockheed Electra 'Lima Alpha' taxies in at Santos Dumont airport.

Below:
A Transbrasil 737-300 passes between two VASP 737s at the Southern end of the terminal building.

Bangkok Booms

Earlier this year an ambitious £1 billion redevelopment programme at Bangkok's Don Muang International Airport was completed. These improvements, and the rapid growth in passenger throughput, have placed Bangkok in the top four aviation hubs in South East Asia, alongside Hong Kong, Singapore and Tokyo. ***Robbie Shaw*** *reports.*

THE FIRST step in the modernisation programme of Don Muang International Airport was the construction of a new runway 3,700m long, paralleling the original overworked runway, which was in dire need of repair. The new runway was completed in 1983 and is served by a new 30m wide parallel taxiway almost 3km long. The old runway is now primarily used by military traffic on the far side of the airfield. Don Muang Air Base is the headquarters of the Royal Thai Air Force, as well as the home of most of the country's military transport fleet. These range from ageing C-47 Dakotas, C-123 Providers, C-130 Hercules to the VIP transport element which includes BAe 748, Boeing 737 and DC-8s. In addition, primary trainers such as the Chipmunk and RFB Fantrainer operate from Don Muang, so runway utilisation is high.

The next step in the programme was the construction in 1985 of a new domestic terminal capable of handling aircraft such as the Airbus A310 operated by the domestic carrier Thai Airways, alongside its Boeing 737s. This terminal is also used for some regional flights, in particular by smaller aircraft such as the F-28s operated by Union of Burma Airways and the Tu-134s of Hang Khong Vietnam. There is, however, ample room at the International Terminal for these flights should it become necessary. During the first full year of operation 1.25 million passengers used the Domestic Terminal.

For many years the old Terminal building struggled to contain the steady build-up of passengers awaiting connecting flights, which pointed to the need for a new

Below:
Bangkok's Don Muang airport is home to Thailand's national airways — Thai Airways which operates an extensive domestic and regional network, and Thai International which looks after the long-haul routes.
All photographs by Robbie Shaw

International Terminal. Designed under a joint Thai-Japanese venture, the new South International Passenger Terminal is the largest and most expensive building to be built in the country. The arrivals hall was the first to open in May 1987, just in time for the closing months of the very successful 'Visit Thailand Year' campaign to increase tourism, and the departures hall following five months later. To give an indication of the new Terminal's capacity, there are 76 check-in desks, as opposed to 40 in the old Terminal, whilst the number of immigration desks and baggage-handling facilities have doubled. The Terminal now has the capacity to handle 16 million passengers a year. To assist the weary travellers with their heavy suitcases, 12 moving walkways are available within the Terminal, along with 23 escalators and 12 lifts.

Within the Terminal, a large 1,500sq m restaurant, and two smaller ones, are all run by the National carrier, Thai International, which is also responsible for the lounge and snack bars as well as the duty-free shops. The airline has a reputation for excellent service, and business class passengers have two Royal Orchid lounges available for relaxation. In addition to toilet and refreshment facilities of a very high standard, the airline has even installed comfortable reclining slumberettes.

For aircraft parking, two 240m long piers were initially built with 15 airbridges; a third one has recently been completed whilst a fourth is currently under construction. On completion there will be 540,000sq m of ramp space available, sufficient for no less than 52 wide-bodied airliners, with 14 of the bridges capable of handling 747s. This is indeed a considerable improvement on the old Terminal, which was equipped with only four airbridges. This meant that there was a considerable number of buses employed carrying passengers across a busy ramp, far from ideal from a flight safety point of view. Happily, this has been remedied since the opening of the new Terminal thanks to the airbridges and moving walkways, buses only being required at the Domestic Terminal which has no airbridges.

At the southern end of the long ramp Thai International has a large maintenance facility, whilst the

Right:
A Kuwait Airways Corporation Boeing 767-200ER on the apron at Bangkok. In the background can be seen the extensive taxiway network serving the two parallel runways.

Centre right:
A Qantas Boeing 747-200 touches down at Don Muang with the control tower and part of the Terminal buildings visible in the background.

Bottom right:
One of three Boeing 737-200s operated by Thai Airways parked on the domestic apron. The airport's International Terminal and its departure gates can be seen at the far end of the apron.

Below:
Bangkok Don Muang International Airport. *British Airways AERAD*

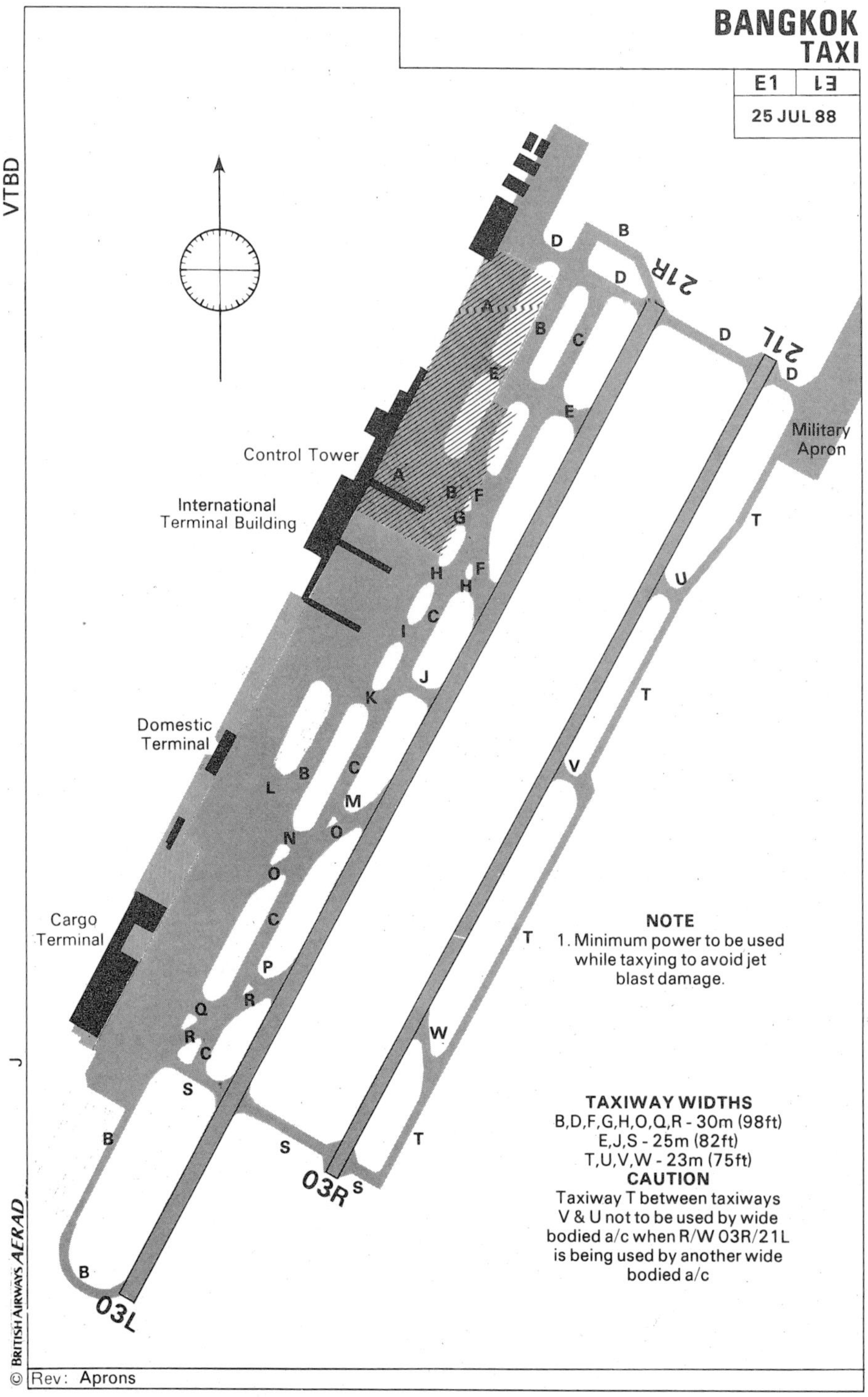

KUWAIT
KUWAIT AIRWAYS
الخطوط الجوية الكويتية

QANTAS
THE SPIRIT OF AUSTRALIA

Thai Airways
HS-TBD

Cargo Village occupies a 57,000sq m site, four times larger than its predecessor, and handles a modest amount of traffic.

Since the opening of the new Terminal, passenger traffic has shown a steady increase. A number of airlines, initially British Airways and Qantas, started to route more of their European-Far East and Australasian traffic through Bangkok, often as an alternative to places like Singapore. One of the main reasons, of course, is the upsurge in tourist traffic, but the modern and efficient facilities at Don Muang must have been a contributing factor. A number of other European airlines are following the British Airways

Above:
Medium and long haul routes are flown by Thai International which is a subsidiary of Thai Airways, both of which are government-owned. Thai International has a fleet which includes 747s, DC-10s and Airbus A300s, one of which is seen on approach to Bangkok.

Below:
Bangladesh Biman is one of the dwindling number of Boeing 707 operators. This 707-320C is a frequent visitor on services from Dhaka.

Right:
Thailand is an increasingly popular destination for tourists from Europe. Unfortunately accidents can happen here, as anywhere, and this Learjet 35A belonging to a German air ambulance company is at Bangkok to evacuate an injured tourist who will be thankful that he took out good insurance cover before he travelled.

Below right:
Hanoi-based Hang Khong Vietnam operates a scheduled service to Bangkok using this Tu-134B — a rare sight to western eyes.

example and increasing the ratio of flights through Bangkok.

Tourists to Bangkok may remember the annoyance of boarding a taxi only to find that it had no meter, and having to haggle with the driver for a reasonable fare. This, too, has been taken care of since the new Terminal was opened. Just inside the exit door of the Terminal is a taxi desk where you simply state your destination, and the price and taxi number is written on your ticket. On leaving the Terminal the taxi driver escorts you to the vehicle, and on reaching your destination you sign the ticket stating that you still have all your luggage safely with you. It is an excellent scheme, and one that a few European airports could emulate.

With the redevelopment now complete, the airport is capable of handling 16 million passengers annually, and it should have sufficient spare capacity to take it into the late 1990s. After that, options include a further development at Don Muang, or alternatively a new airport to be constructed at Nong Ngu Hao.

There is no doubt that the majority, if not all, of the 46 airlines currently serving Bangkok are happy with the facilities available. Thai International, the National flag carrier, was responsible for about 30% of traffic through the airport. However, early in 1988 it merged with the largest of Thailand's domestic carriers, Thai Airways, which is based at Don Muang with a fleet of A310 Airbus and Boeing 737s.

Boeing's 747 — The Airliner

That Changed the World

*The year 1989 saw the first deliveries of Boeing's new 747-400 series to airline customers. Twenty years and over 800 orders after its inaugural flight, **Leo Marriott** looks back over the Jumbo's career, and forward to its latest incarnation — the 400.*

THE FIRST flight of any aircraft is a heartstopping moment for all involved in its design, development and production, but in the case of the Boeing 747 those concerned must have realised that they were watching one of the most significant events in the history of civil aviation. By far the largest civil airliner ever built, more than twice the weight and capacity of any other commercial aircraft flying at the time, the sight of the 700,000lb aeroplane lifting easily off the runway at Boeing's Everett airfield was positively awe-inspiring. Yet today the arrival or departure of one of these aircraft is an everyday occurrence, although heads still turn in disbelief at the thought of such a giant structure getting airborne. In fact, not only does it get airborne, but the Boeing 747 has proved to be one of the world's safest and most successful aircraft.

In the postwar period air transport expanded only gradually using piston-engined equipment and it was not until the widespread introduction of the jet airliner, such as the Boeing 707 and redesigned Comet, that long-range air travel really became affordable to a wide range of people. The comfort and speed of the jets stimulated demand to an extent which had not been fully anticipated and for a while airlines made do with increasing the seating capacity of existing types. The ultimate result of this trend was the DC-8-60 series which could seat around 250 passengers in a much stretched fuselage. This version of the DC-8 first flew in 1966 and proved very popular with airlines, some 260 eventually being built.

Although Boeing was aware of the demand for extra capacity, the Boeing 707 was structurally unsuitable for stretching along the lines of the DC-8. The company was therefore faced with the development of an all-new aircraft if it was to remain in a competitive position. As far back as 1963 an engineering planning group had been set up to look at possible designs which would meet the anticipated passenger and cargo demands of the 1970s. A year later Boeing was engaged in a head-on competition with Lockheed and Douglas to build a massive new military freighter. This was won by Lockheed, which went on to build the C-5A Galaxy, but the loss of this spurred Boeing on to look at the possibility of building a really large airliner which would significantly lower the seat/mile costs of operation. Serious design studies commenced in 1965 and included enlarged versions of the 707 with a stretched or double-deck fuselage.

One unknown factor which had to be allowed for was the possible success of supersonic aircraft such as the Anglo-French Concorde or the American SST. It was possible that, in the future, this type of aircraft would soak up all the passenger traffic leaving Boeing's new project with unfilled seats. From the start, therefore, the new Boeing 747 was also intended to be able to serve as a freighter and this really determined the aircraft's size and shape. A cargo deck which would carry 8ft wide containers two abreast determined the fuselage width, and a requirement for unobstructed nose loading resulted in the flightdeck being raised up in a distinctive bulge above the forward fuselage.

Although Boeing was convinced that the project was technically feasible, there remained the problem of convincing the airlines that such a huge aircraft would be a commercial success. The breakthrough was not long in coming when Pan American's president, Juan Trippe, announced a $20 million order for 25 Boeing 747s in April 1966. With an order from this influential airline safely signed, the rest of the world's majors were quick to follow and a total of 85 aircraft were ordered that year. At this stage the airlines involved began to play an active part in the refinement of the design — in particular the configuration of the cabin layout and seating arrangement. This resulted in the 10-abreast rows, split up by twin aisles, which is today standard on most wide-body aircraft. Below the main cabin was over 6,000cu ft of cargo space of which just over half was allocated for passenger baggage, but the remainder provided a significant revenue earning facility for general air freight — a factor which has contributed to the aircraft's commercial success.

In order to build the world's largest airliner, Boeing had first to build the world's largest building measured by volume (200 million cu ft) in order to house the production lines. It is perhaps this aspect of the programme which illustrates the enormous capability of American heavy industry and shows why smaller countries such as Britain and France can only tackle the world market by pooling their resources. Clearance of the 780-acre site at Everett, Washington, began in mid-1966 and, incredibly, production work on the prototype began only six months later. In fact the factory and aircraft building programme were closely related, with Boeing's workers moving into sections of the factory and other buildings as they were completed. Those involved recall an air of exhilarating bedlam as the production lines took shape while construction crews hammered, welded and riveted only yards away in an adjacent bay.

Of course, even a giant company such as Boeing could not build a 747 unaided: more than 60% of work on the aircraft's structure and systems was subcontracted to over 1,500 prime suppliers, and no less than

Left:
The scene at Boeing's Everett facility on 30 September 1968, when the prototype 747 was rolled out. *Boeing*

Below:
HB-IGD is one of two 747-357 aircraft flown by Swissair. The airline also operates three 300 series Combi models. *Robbie Shaw*

Bottom:
United Airlines has the world's largest fleet of the Boeing 747SP (11 currently in service). This version is easily recognised by its shortened fuselage and enlarged tailfin, and is optimised for long range routes. *Robbie Shaw*

Below right:
The various cargo versions of the Boeing 747 have accounted for a significant number of orders. This 747-200F is one of eight flown by Northwest Airlines which was also the launch customer for the latest 400 model. *Robbie Shaw*

Bottom right:
UTA is France's largest independent airline and, apart from this single 200F, it also operates four 747-300s with two 400s on order. *Robbie Shaw*

15,000 secondary sources. These were located in every one of America's states as well as six other countries, and the Everett complex includes its own railhead to receive the constant stream of parts which can range from small packs of rivets to complete fuselage or wing sections.

Despite the complexity of the programme, and the inevitable design changes, the prototype Boeing 747 was rolled out to the accompaniment of bands, cheers and speeches on 30 September 1968

— one day ahead of the schedule set over two years previously.

However, there remained a considerable amount of work and ground testing to be carried out before the first flight was made on Sunday 9 February 1969. The pilot on this momentous occasion was Jack Wadell, assisted by his co-pilot Brien Wygle and flight engineer Jess Wallick. All three had been associated with Boeing or the 747 project for some time and were thoroughly familiar with the aircraft and its systems. Although the weather on that Sunday was marginal for the important first flight — cold and cloudy with snow patches still lying — the big jet eventually took off before a large gathering of Boeing workers at 11.34am and remained in the air for over an hour. There was a minor problem with the trailing edge flaps and the undercarriage was not retracted, but this was cured for the second flight when the aircraft flew in the 'cleaned-up' state.

Within a few days of the first flight the second 747 was rolled out, to be followed by three more by the end of May 1969. These five aircraft (Boeing's prototype, two Pan Am and two TWA) shared the complex test programme which lasted 11 months and required 1,400 flight hours at a cost of $28 million. In addition, two complete airframes were used for the static ground testing programme which continued well after FAA certification which was achieved on 30 December 1969.

The early 747 models were powered exclusively by the Pratt &

Whitney JT9D-3 which offered 43,500lb thrust to cope with the aircraft's gross weight of 710,000lb — an increase of 30,000lb from the design target of 680,000lb. One of the features of the 747 is the way in which it has been a forcing ground for engine development with heavier and longer range versions demanding more power, while the soaring cost of fuel in the early 1970s led to major advances in efficiency. Although the JT9D has been gradually improved to give over 55,000lb thrust, other engine manufacturers vied to hang their products on the 747. In 1973 the original prototype 747 was re-engined with General Electric CF6-50D engines, each giving 51,000lb thrust, but it was not until the end of 1975 that the first civil versions powered by this engine were delivered. These were 747-200B Combis for KLM, although the CF6 had in the meantime been selected as the standard powerplant for the military E4 version.

Rolls-Royce was the third engine manufacturer to power the 747 but an RB211-powered model did not fly until 1976 (when it made up for lost time by achieving a world record take-off weight of 840,500lb on 1 November 1976). Although the initial order for Rolls-Royce-powered aircraft was predictably from British Airways, several other operators were quick to follow, notably Qantas and Singapore Airlines, and the engine has achieved an excellent reputation.

Commercial services with the 747 commenced with Pan Am in January 1970, followed by TWA the next month. The first European airline to introduce the type was Lufthansa (April 1970) and Air France was next in the following June. British operations commenced in April 1971 in BOAC colours as British Airways was not formed until 1973 when new aircraft were delivered in the revised livery and the old examples repainted. Since those early days, a total of 910 Boeing 747 variants have been ordered (as at February 1989) and no less than 684 were actually in service on 31 August 1988. The aircraft has been responsible for a major revolution in air transport offering significant reductions in seat/mile costs. This has meant that long-range air travel has become available to millions of people around the world, materially increasing opportunities for trade, tourism, and international contact. Even today, the aircraft has no direct rival in the field of long-range mass transportation although several types such as the Airbus A330/340 and the MD-11 are nibbling at the fringes of the market.

Compared to earlier jets such as the Boeing 707 and the DC-8, the 747 design has only undergone gradual evolutionary changes and most models retain the aerodynamic form and overall dimensions of the original 100 series model — a tribute to the soundness of the original design. Most changes have

been directed at increasing the aircraft's operating weight to allow increased range and payload, and the incorporation of alternative power plants. In accordance with the original concept, several freight versions have been produced, while the latest version, the 747-400, takes advantage of modern aerodynamic and electronic refinements. A complete list of civil 747 variants to date is as follows:

747-100 Original production version with Pratt & Whitney JT9D-1 or 3 engines. 738,000lb max weight.

747-100B Improved version with strengthened structure and undercarriage to allow operation at up to 753,000lb max weight. Choice of JT9D, CF6 or RB211 engines. Introduced in 1978.

747-200B Long-range model with max weights up to 833,000lb. Available with uprated JT9D-7, CF6-50E and RB211-524B/C/D engines. First flight 11 October 1970.

747-200B Combi Based on the standard 200B, but features a side-loading cargo door aft of the wing and can be operated as an all-passenger aircraft or in a mixed cargo/passenger configuration; pallets or containers are carried in the after half of the cabin which has a specially strengthened floor. Performance, dimensions and weights identical to the standard 200B.

747-200C Similar to the Combi except that the whole cabin floor is strengthened for freight and consequently can be used in the all-cargo role if required. A cargo-handling system is installed to reduce dependence on airport facilities.

747-200F A dedicated all-cargo version of the 747 with an upward hinging nose. Can carry a 248,000lb payload over 4,440 miles or a 127,940lb load over 7,410 miles.

747SP (Special Performance). The only 747 version with a significant variation in external dimensions. Fuselage length reduced by 48ft. Other structural changes result in gross weights of up to 703,000lb, a fuel capacity of 41,930gal and ranges up to 8,880 miles. Intended for ultra-long ranges, it has been eclipsed recently by improved performance from standard models, and less than 40 are in service. First flight 4 July 1975.

747-100SR A version of the 100B optimised for short-range high-frequency services. First flight 4 September 1973.

Below left:
Powered by four General Electric CF6-50E2 turbofans, the Boeing 747-200F's main cargo deck can be loaded through both the nose loading door and the side cargo door at the rear. *Lufthansa*

Diagram of 747-400 flightdeck layout. *Boeing*

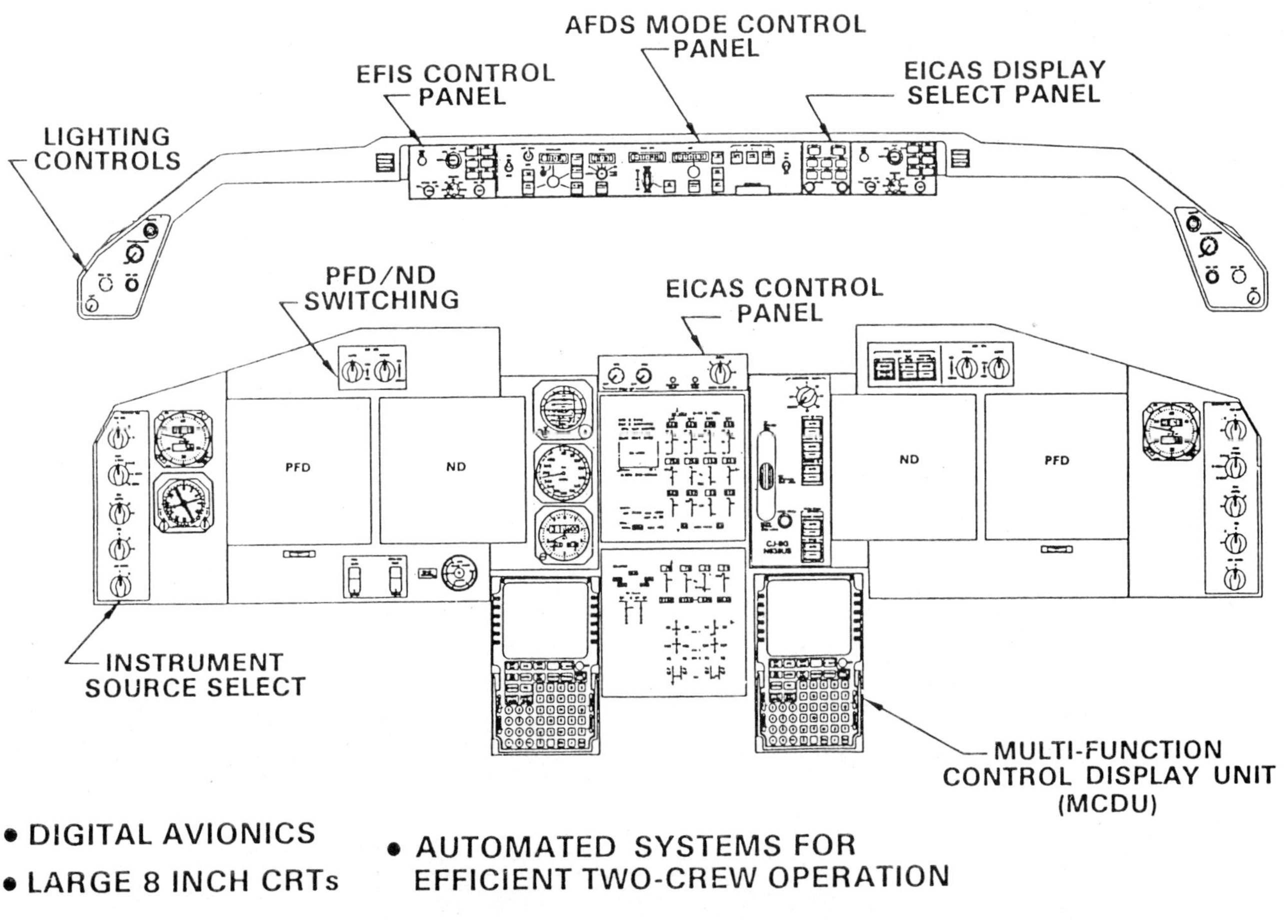

Above:
The advanced flightdeck of the 747-400 is dominated by the six CRT screens which replace most of the conventional instrumentation. Earlier 747 models had 971 switches, lights and instruments in the cockpit; this has been reduced to 365 on the 400. This simplification is a major factor in certificating the aircraft for a two-pilot crew. *Boeing*

Above right:
Boeing's 747-400 is the latest variant to fly; this artist's impression shows the first of six PW4056-powered aircraft ordered by Korean Air in April 1988. The performance of this variant will allow the airline to carry a full passenger and cargo payload nonstop from Seoul to New York. *Boeing*

Below:
TWA was the second customer for the 747 and started transatlantic services in March 1970. It currently flies 14 747-100s one of which, N133TW, is shown at Heathrow. *Robbie Shaw*

Below right:
The 747-300 features an extended upper deck which increases the maximum passenger load to 660 seats and it has proved popular, with 80 aircraft ordered. This example (HZ-AIO) is one of 10 belonging to Saudi Arabian Airlines.
W. A. Blanchard

747-300 Basically similar to the 200B except for the introduction of a stretched upper deck which increases maximum seating capacity from 550 to 660, although most airlines operate mixed class layouts utilising less seats. Gross weights and performance identical to the 200B. The upper deck extension can be applied to earlier 100B and 200B models which are then redesignated as 747-300.

747-400 Latest variant externally identified by the addition of vertical winglets on the extended wingtips and the stretched upper deck as on the earlier 300. Maximum weight is increased to 873,000lb and a choice of PW4256, CF6-80C2 and Rolls-Royce RB211-524D4D engines in

the 56-58,000lb thrust class is on offer. First flight 29 April 1988.

Although superficially similar to the 300, the current 400 version is almost a new aircraft internally. Despite the aircraft's legendary strength and safety record, Boeing have beefed up the airframe and improved corrosion treatment in known trouble areas. New generation powerplants from each of the major manufacturers offer increased thrust and improved reliability. All three feature digital electronic fuel control but will ultimately have full authority digital engine control. These techniques result in a total fuel burn around 5-10% less than the earlier versions.

However, it is on the flightdeck that most of the changes become apparent. Despite the size of the 747, the flightdeck has always seemed somewhat cramped compared to other contemporary airliners such as the DC-10 or TriStar. The layout has now been improved and six large 'glass' screens now display most of the information which previously needed serried ranks of conventional instruments. Each pilot has two screens, a Primary Flight Display (PFD) and a Navigation Display (ND). The PFD shows the horizon and attitude as well as airspeed, altitude and climb/descent information. The ND is basically an electronic horizontal situation indicator (Ehsi) with the aircraft symbol in the centre and capable of operating in four different modes. The two screens are interchangeable — for instance the PFD can be transferred to the ND screen if the original PFD should fail and the other pilot can then reselect his screens as required.

Stacked vertically in the centre of the panel are two engine indication and crew alerting system (Eicas) screens. The upper screen displays primary engine data such as fan speeds and turbine temperatures, while the lower can show the functioning of various systems such as electrical power supply or hydraulics and can be used for monitoring operation and to highlight faults.

Other computers include a flight management system which controls engine, power and fuel flow settings as well as determining vertical and horizontal flight profiles. Also fitted is a Central Maintenance Computer (CMC) which can be used by ground engineers to locate system and engine faults as well as by the pilots, but only above 10,000ft so as to avoid unnecessary distractions in the busy departure and arrival phases.

All these changes have led to a reduction in the crew requirement and the aircraft is designed to be operated by only two pilots with no flight engineer. Despite this, the long ranges of which the aircraft is capable mean that pilots will exceed their permitted duty hours and consequently a small 'bedroom' with two bunks is situated immediately behind the flightdeck to enable relief crew members to be carried. A similar arrangement is incorporated right aft where an optional upstairs loft area provides rest facilities for cabin crew. Other improvements in the cabin include restyling with bigger baggage lockers, much improved fire resistance of the cabin furnishing materials, improved ventilation and provision for the latest generation of in-flight entertainment devices including seatback videos.

In its 20-year lifespan, the Boeing 747 has flown over 800 million passengers. The latest models will take the aircraft well into the 21st century and in prospect are enlarged models with full-length upper decks and redesigned wings. No doubt the advances in aerospace technology will offer the chance for even more sophisticated control and propulsion systems. Having already secured its place in aviation history, the Boeing 747 will be making history for many decades to come.

Airbus — International Success Story

The European Airbus Programme is an international and technological success story. ***Philip J. Birtles*** *investigates.*

FORMED initially in December 1970 by the British, French and German Governments with their national aerospace industries, the major aim of the European Airbus Programme was to produce a comprehensive and, in the long term, profitable range of world-beating airliners to compete strongly with the massive US West Coast airliner suppliers who have a large domestic market with the airlines and lucrative US Government contracts. To start such an undertaking takes many years of investment, market credibility and engineering development, even using the existing manufacturing facilities of the member companies. The programme is so vast that only governments can underwrite with long-term loans the development programme which could span 15 years before a return on investment could be seen.

Following individual design studies by the national aircraft industries, it was soon realised that to spread the overall development costs and at the same time increase the security of the combined home market, the programme would require international collaboration. Studies centred around a 300-seat twin-aisle wide-bodied twin-jet, short-to-medium range airliner, to provide quiet, fuel-efficient air transport on the high density routes of the world. These studies centred around the HBN-100, an Anglo-French consortium of Hawker Siddeley Aviation, Breguet and Nord Aviation. Meanwhile Sud Aviation, later to become Aérospatiale, was undertaking its own studies, and in Britain, BAC was proposing wide-

Below:
Airbus A300 wing assembly at BAe Chester. *British Aerospace (BAe)*

Top:
Re-registered F-OCAZ, the prototype embarked on a major sales campaign and is shown at Dublin in 1974 while a sales presentation was made to Aer Lingus.
P. Birtles

Above:
The short-to-medium range A310 is shorter than the A300 but retains the same fuselage cross-section. This example operated by Swissair is seen at Zürich.
P. Birtles

bodied, rear-engined twin-jets in the form of the BAC 2-11 and the larger BAC 3-11.

In the event, the French Government chose Aérospatiale to be the national lead in the Airbus Programme, and Germany joined the team in the form of Deutsche Airbus, consisting mainly of MBB, the remaining minor partners being CASA of Spain and Fokker of Holland. Design and development work commenced in September 1967 on what was known as the A300 with a choice of suitable newly-developed high by-pass engines from Rolls-Royce, General Electric and Pratt & Whitney. As a result of studies during the project definition, the aircraft was reduced in size to become known as the A300B. The engines chosen initially were the GE CF-6-50, similar to those used on the McDonnell Douglas DC-10, to allow some commonality with the airline's engine maintenance. Later the Pratt & Whitney JT9D was also chosen, but Rolls-Royce, like British Airways as customers, has been a long time joining such a widely acclaimed and popular programme.

An early setback to the Airbus was a stipulation by the British Government of the day, that for prototype construction to commence, a total of 75 A300s should be on firm order by the three national airlines of Britain, France and Germany. This was clearly a totally unreasonable condition so early in the programme, but when the orders were not received, the British Government withdrew its support in March 1969. However, it is to the credit of Hawker Siddeley Aviation that it remained in the consortium, financing it with its own capital, and shouldering a share of about 12%. Britain retained responsibility for the design, development and production of the wing, creating in the process a new science in aerodynamics with the advanced profile which reduced drag and weight while increasing fuel capacity and overall simplicity. This advanced technology has now been adopted worldwide as the standard in fuel-efficient aircraft.

Hawker Siddeley was not only wing producer, but key employees were seconded to Airbus Industrie at Toulouse as part of the administration, sales, training and product support organisation, a role which is still undertaken today by British Aerospace. The factories of Hawker Siddeley Aviation shared the work on the Airbus wings with Hatfield (being the design centre), and final assembly at Chester. Changes in more recent years with British Aerospace have seen the design responsibility moved to Bristol, and while assembly of the major wing structure of all Airbus types is still in the vast assembly hall at Broughton, near Chester, equipping the A320 wings and fitting of control surfaces is undertaken at Filton.

Aérospatiale is responsible for the assembly of the nose and final assembly at Toulouse, while the fuselage centre-section is manufactured at Nantes. The contribution from MBB is the remainder of the fuselage sections, equipping of the A300 and A310 wings and final furnishing to the customer requirements. Fokker supplies all the wing-moving control surfaces and CASA is responsible for the horizon-

Above:
Philippine Airlines is government-owned and has six of the B4 variant.
Robbie Shaw

Below:
Airbus Industrie has had particular success in the Far East market. Malaysian Airlines operate four A300B4s. *Robbie Shaw*

Bottom:
Air India operates both A300s and A310s. This is one of their A310s at Hong Kong.
Robbie Shaw

Above:
Results of the flight-test programme proved the A310 to be lighter, and able to fly higher, faster and further than predicted. This example is operated by Turkish Airlines and is photographed at Ercan State Airport in northern Cyprus. *Leo Marriott*

Right:
Airbus Industrie is heavily committed to the use of the latest technology in its aircraft. An example of this is the winglets originally developed for the A300B-600 but now used on all Airbus types. *Robbie Shaw*

Below:
The latest Airbus product in service is the A320 which competes with the Boeing 737. This example is the first with IAE V2500 engines and was shown at Farnborough in 1988. *Leo Marriott*

American
AIR INTER
A320
aerospatiale

tal tail surfaces and access doors. Other countries subsequently involved with the Airbus programme include Belgium, Australia, Canada, and a substantial contribution from the USA in the form of engines, equipment and raw materials. Due to the initial lack of support of the British Government, very little British equipment was selected for the early aircraft.

Initial sales were slow, as was to be expected with a new organisation which had to gain credibility and respect for itself and its products. Early orders were from Air France and a reluctant Lufthansa, now a major operator, followed by a gradual acceptance by other major European national carriers.

Following the first flight of the prototype A300-B1 on 28 October 1972 (of which only two were built, being followed by the A300-B2 as the improved production version) the development programmes went so smoothly that, before certification in March 1974, the first aircraft was able to participate in demonstration tours worldwide. The confidence gained resulted in modest sales in South America and Africa, but opened up major markets in the Middle East, India and Asia. The Airbus is used extensively throughout South East Asia and the surrounding countries, including China. A breakthrough in the USA was to win a major order from Eastern Airlines in a highly com-

Left:
This picture of the flight test centre at Toulouse gives some idea of the scale of the Airbus programme. Note the British Airways A320 to the left of the building.
Airbus Industrie

Below left:
A320s on final assembly: a production rate of eight aircraft per month is planned for the 1990s. The centre aircraft is the first of 100 for Northwest Airlines.
Airbus Industrie

Below:
Introduced into Lufthansa service during 1987, the A300-600 is an advanced version of the A300 which operates on short and medium haul routes. *Lufthansa*

petitive sales battle. Since then other significant sales have been made to Pan Am, Continental and American Airlines.

The policy of product development continued with the similar but increased range A300-B4, while various other projects were being studied — the most promising being the A300-B10. This became the A310 when the go-ahead was given in July 1978 for an aircraft similar to the A300, but with a smaller capacity fuselage and new wing design. This aircraft received launch orders from Lufthansa in April 1979 and Swissair, which set the scene for sales of this aircraft to both existing and new customers.

By this time — through the foresight of Hawker Siddeley — British Aerospace were able to become a full partner on 1 January 1979 with a 20% share in Airbus Industrie, through the renewed support of the British Government, who backed the A310. New technology introduced on the A310 included further improvements in wing aerodynamics, which has now changed from an art into a science, greater use of composite materials including such vital areas as the fin, and CRT displays in the cockpit replacing the old analogue instruments and reducing the workload sufficiently to allow two-man operation. Extended range versions of the A310 featured drag-reducing winglets. The first A310 made its maiden flight on 3 April 1982 and entered service with Lufthansa one year and six days later.

Much of the new technology developed on the A310 was adapted to the A300B airframe to become the updated A300-600. In addition to all the equipment refinements, the rear fuselage was adapted to the A310 shape, giving additional seating capacity. An unusual feature of the A300 and A310 is the gradual upward slope of the rear cabin floor, following the line of fuselage maximum width. Not only does this optimise the passenger accommodation, but also leaves room for underfloor baggage and freight in the standard airline containers, allowing compatibility with the other long range tri-jet airliners and the Boeing 747.

The next major step in the development programme was to produce a 150-seater. After difficult battles with the respective governments for launching aid — particularly, once again, the British Government — approval was given in March 1984 for what has become the all-new A320. It has a growth capability to 180 seats and like the Airbus variants before, is well ahead of the competition in the new generation of airliners. The only apparent threat in technological terms was the unducted fan-powered Boeing 757 or MD-80 variants, but Boeing shelved its project in favour of 737 developments, and McDonnell Douglas has yet to achieve an order at the time of writing.

The A320 is the first narrow-bodied Airbus aircraft, and was a resounding success from the start, with some 600 orders before the maiden flight. Launch orders were from Air France and British Caledonian, which was taken over by British Airways before the delivery

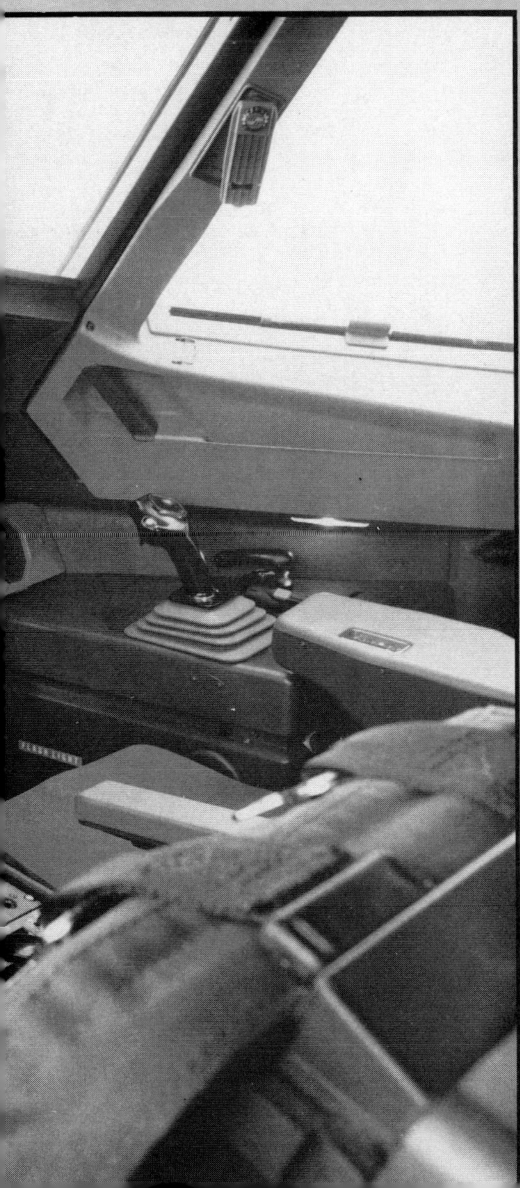

Above:
The future of Airbus depends on the success of the twin-engined A330 and four-engined A340. The former is illustrated here and will seat up to 335 passengers to compete with aircraft such as the MD-11. *BAe*

Left:
The A320 is the first 'fly-by-wire' civil airliner and the cockpit features sidestick controllers and a 'glass' cockpit. *Airbus Industrie*

of the first aircraft. Despite some rattling of the cage, British Airways accepted its first aircraft, somewhat reluctantly, having been the only major European airline not to see a need for one of the Airbus products, but are reportedly delighted with the performance and support of the aircraft. However, recent orders have confirmed BA's long-term loyalty to Boeing products and further Airbus orders appear unlikely.

Amongst new technology adopted by the A320 is fly-by-wire controls, side-stick control columns and a fully computerised flight system which together make the aircraft inherently safer than earlier generation aircraft.

Safety is always an emotive word, but the Airbus aircraft proved in service over the years to be safe and rugged. One aircraft had a bomb explode in the rear toilet, and despite severe damage it landed safely with no further injuries to the passengers. An Airbus was destroyed during crew training with Egyptair, killing the crew, an Iranian Airbus was shot down by missiles from an American warship and an early A320 delivered to Air France crashed at an airshow when flown into trees. Despite having a cabin full of passengers, only three were killed, although the aircraft eventually became a totally burnt-out wreck. Investigations into these accidents put no blame on to the aircraft or its systems.

Work has already started on the next members of the growing Airbus family. These are the A330 and A340 formally launched in June 1987, which share the same wing and a similar fuselage, but which varies in length. The A330 is a high-capacity regional twin-jet which carries 335 to 440 passengers up to 4,600nm. The A340 is offered for the long thin routes and will be powered by four CFM56 engines each developing 31,200lb thrust. The two versions of the A340 offered

Above:
Powered by four CFM-56-5C-2 engines, the 295-seat A340 shares the same fuselage and many other common components with the A330, but is intended for long range routes which will not support larger aircraft. *Airbus Industrie*

initially will be able to carry from 262 to 440 passengers over ranges of up to 7,450nm with the lower number of passengers. By mid-1988, 149 commitments had been received from 12 customers, in stiff competition with the McDonnell Douglas MD-11, for the pair of new airliners are expected to enter service from 1992 onwards. A recent rash of orders has firmly established both types, and in April 1989 Cathay Pacific ordered 10 A330s powered by Rolls-Royce RB211-524L engines. A significant breakthrough for the British aero engine manufacturer.

For many years Boeing has maintained a monopoly in the high-density long range jet airliner market with their 747 turbojet. To compete with this market requires a great deal of investment with a long return to profitability. However, Airbus Industrie and McDonnell Douglas have conceived a cost-effective response by combining an extended MD-11 fuselage with the new technology wing of the Airbus A330 produced by British Aerospace. This will offer a highly competitive high-density, long range tri-jet to challenge the leadership of the Boeing 747 with the world's airlines, but a final go-ahead will depend on the results of further talks between Airbus and McDonnell Douglas.

Airbus Industrie is now a well established organisation with a viable and commercial family of airlines to meet the needs of the world's major airliners into the next century. By its success it has shown that Europe has the ability to meet the competition from the industrial might of the United States — and to win! As such it is a shining example of what European companies — not solely in the aerospace industry — might achieve in the years to come.

Flying Over the Roof of the World

Very occasionally an airline discovers that only one aircraft type will fit its operational needs. One such airline is the Indian regional operator Vayudoot. ***Rolf Christ*** *reports from the roof of the world.*

Above:
The Dornier Do228 design applies high technology to the small commuter airliner. Design work on the advanced TNT wing was sponsored by the German Federal Ministry of Research and Technology. *Dornier 39/141969/30a*

ONE OF THE features of today's airline scene is the tremendous variety of aircraft types designed for use on commuter and regional short haul routes. At first glance the choice available to an operator is overwhelming, but in practice each aircraft type offers a particular blend of capabilities and operating economics which narrows the final choice of aircraft to only two or three types. Occasionally a situation arises where an airline has such unusual circumstances that only one aircraft will fit the bill. A case in point is the Indian regional airline Vayudoot, formed in 1981 to provide a network of services in northwest and eastern India.

Although the airline now operates a number of Fokker F27 and HAL 748s (licence-built BAe 748), the bulk of its fleet is made up of 10 Dornier Do228 twin-engined 19-seat airliners. This aircraft first flew in 1981 and is now also produced in India. Powered by two Garrett 7PE 331-5 turboprops, each producing 715shp, the aircraft features the advanced technology TNT wing developed by Dornier which endows it with an excellent short field performance coupled with a high cruising speed (231kt). Apart from size, the clinching factor in Vayudoot's selection of the Dornier was the aircraft's performance out of high altitude airfields such as those in the Kashmir region of India. The problems encountered in this type of flying are described by Rolf Christ, one of Dornier's marketing executives:

'Kashmir, the Himalayan district of the upper Indus in northwest India, is one of the country's most popular tourist regions. Consequently, Vayudoot has pressed ahead with the development of air services to this region using the Dornier Do228 light transport aircraft. Seating 19 passengers, the aircraft's characteristics of good manoeuvrability, excellent take-off and landing performance and high rate of climb make it ideal for this task.

'Initial evaluation flights with the Dornier were carried out in the summer of 1987 using crews with considerable experience of flying in this mountainous area. Proving and certification flights were made to the airports at Shimla, Kulu and Kargil with Dornier's chief test pilot Dieter Thomas involved in the latter stages of the evaluation.

'Shimla is an entirely new airport situated astride a mountain range in the Himalayan foothills, some 20km from Jubberhati. During the construction of the 3,300ft (1,000m) runway at an altitude of 5,500ft (1,700m) three mountain tops had to be removed with bulldozers. At one end of the runway a further hillcroot rises 90ft so that landings are only possible from the west and the take-off is then made in the opposite direction assisted by the steep slope.

Above:
Operation of the Do228 from the mountain strip at Shimla is obviously a major spectator sport! *Dornier 15/145154*

Below:
This aerial view of the strip at Shimla shows the precarious nature of this location. The runway is situated on the flattened top of a 5,500ft peak. Visible in the foreground is the remaining peak which forms a major obstruction for landings and take-offs. *Dornier 15/145147*

'The runway at Shimla can be compared with the deck of an aircraft carrier, except that it cannot be turned into wind! Constant observation of the windsock is necessary and crews need to be aware of the possibility of windshear caused by the wind blowing over the hilltops. An additional problem is that, even at this altitude, the temperatures can exceed 30°C in summer.

'Despite the demands of this difficult site, the aircraft coped well in trials and Vayudoot was able to

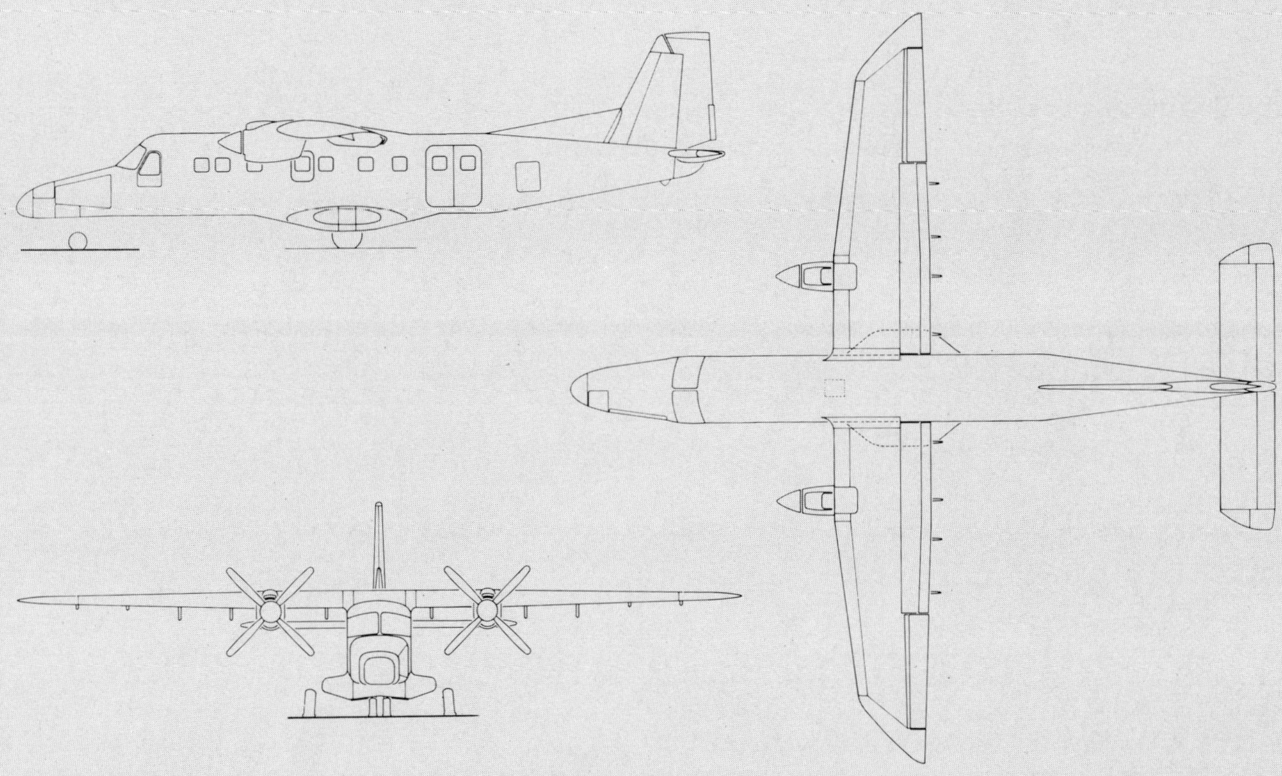

Do228 three-view drawing. *Dornier 25/133469*

commence a twice-daily scheduled flight from Delhi to Shimla in the summer of 1988. This flight makes a stop-over at Chandigarh with a flight time of 55min on the sector followed by a 35min flight to Shimla. On four days a week an onward connection is flown to the strip at Kulu, 150km north of Shimla. Apart from these scheduled links, Kulu is also used as a base for tourist flights through the Himalaya range. These are also operated by Vayudoot Do228s giving visitors a breathtaking experience and an unequalled opportunity to appreciate the majesty of the world's highest mountain range.

'Kargil airport in the northwest edge of Kashmir offers a similar attraction and is situated in the centre of a valley surrounded by mountains rising to between 12,000ft and 15,000ft (4,000m to 5,000m). This airport, with its sand and gravel runway, can be reached from the region's capital, Srinagar, only under visual conditions involving flights along rugged mountain valleys at altitudes up to almost 12,000ft. There are no radio navigation aids and the flight through narrow ravines requires a very manoeuvrable aircraft together with flight crews familiar with the unique problems of mountain flying. Occasionally, when weather conditions deteriorate unexpectedly, it can be necessary to turn within the confines of a valley and return to the original departure airfield. In these circumstances Vayudoot pilots are recommended to fly at reduced airspeed (150-120kt) with propeller rpm set at 100% and 5° of flap selected.

'Landing at Kargil is not quite as dramatic as at Shimla since the well prepared sandy runway is 9,000ft (2,750m) long, although the airfield elevation is 9,200ft (2,800m) and temperatures also exceed 30°C in the summer. In the winter, the strip is closed due to heavy snow coverage.

'Despite the difficulties of operating in such hostile terrain, Vayudoot has succeeded in maintaining a safe and reliable service. This is undoubtedly as much due to the characteristics of the Do228 as it is to the professionalism of the airline's crews.'

The experience of Vayudoot is echoed by many airlines around the world which have ordered over 150 Do228s to date in addition to another 150 which will be built on the Hindustan Aeronautics production line in India. Building on this success, the company announced on 3 August 1988 that production of the 30-seat Do328 will go ahead with first deliveries scheduled for 1992. This aircraft retains the advanced TNT wing design but will offer substantial performance increases as well as extra capacity and, together with the Do228, will enable the company to offer a comprehensive product range in the next decade.

(The author gratefully acknowledges the co-operation of Dornier GmbH in the preparation of this article.)

Paramount's MD-83

For many years the workhorse of the British inclusive tour market has been the Boeing 737 but in recent years a new contender has appeared in the shape of the McDonnell Douglas MD-83, a derivative of the well known DC-9.
***Leo Marriott** accompanies Capt Dave Hipperson and First Officer 'Dusty' Rhodes on a Paramount MD-83 flight to Crete.*

Above:
Welcome Aboard! Paramount crews are proud of their MD-83s which are kept immaculately clean — inside and out.

Above right:
The MD-83 flightdeck follows a conventional layout.

Right:
The roof panel houses controls for electrical power distribution, the APU, icing protection, cabin air conditioning and other ancillary systems. The master warning panel with its illuminated captions is at the lower edge.

McDONNELL Douglas's DC-9 first flew in 1965 and was produced in several variants with a total of 976 aircraft delivered (including the C-9 military version but excluding the MD-80 variants). During the 1970s, rising fuel costs led to demands from the airlines for greater passenger capacity in order to increase efficiency. The result was the DC-9-80 series (also known as the Super 80) which incorporated a fuselage stretch of 14ft 5in and a 28% increase in wing area compared to the previous series 50 model. Other less obvious changes included uprated and more efficient Pratt & Whitney JT8D-200 engines, increased fuel capacity in the larger wing, redesigned cabin interior and new electronic control and instrumentation systems.

The first Super 80 flew on 18 October 1979 and the type entered airline service with the launch customer, Swissair, in September the following year. In 1983 the McDonnell Douglas company adopted a new system of designations for its commercial aircraft and the Super 80 and its derivatives became known as the MD-80 series. Apart from the basic MD-81 and the 'hot and high' MD-82, one of the most popular versions is the MD-83 which, with extra fuel in new cargo bay tanks and 21,000lb thrust JT8D-219 engines, offers increased range and payload with a maximum take-off weight of 160,000lb. It is interesting to compare this with the

77,700lb gross weight of the original DC-9-10. To the end of 1988 over 1,200 MD-80 series aircraft have been ordered, including 25 being co-produced in China.

Despite the international popularity of the DC-9 and its successors, the type has had little impact on the British market. In the early days operators preferred the home-built BAC One-Eleven and difficulties with certifying the DC-9 in the UK (mainly due to the differing philosophies of the respective airworthiness authorities, rather than any intrinsic problems with the aircraft) have meant that only one British airline, British Midland, has ever operated the type. Currently, BMA have 10 DC-9s and is well pleased with their performance. The MD-80 series has also been eclipsed on the British market, most potential customers going for the lower capacity Boeing 737-300 or the larger 757.

Given this background, it was surprising that the first British order for MD-80 series aircraft should come from a brand-new airline — Paramount Airways. This Bristol-based airline was formed in the summer of 1986 and started operations on 1 May 1987 with two brand-new MD-83 aircraft. Aiming

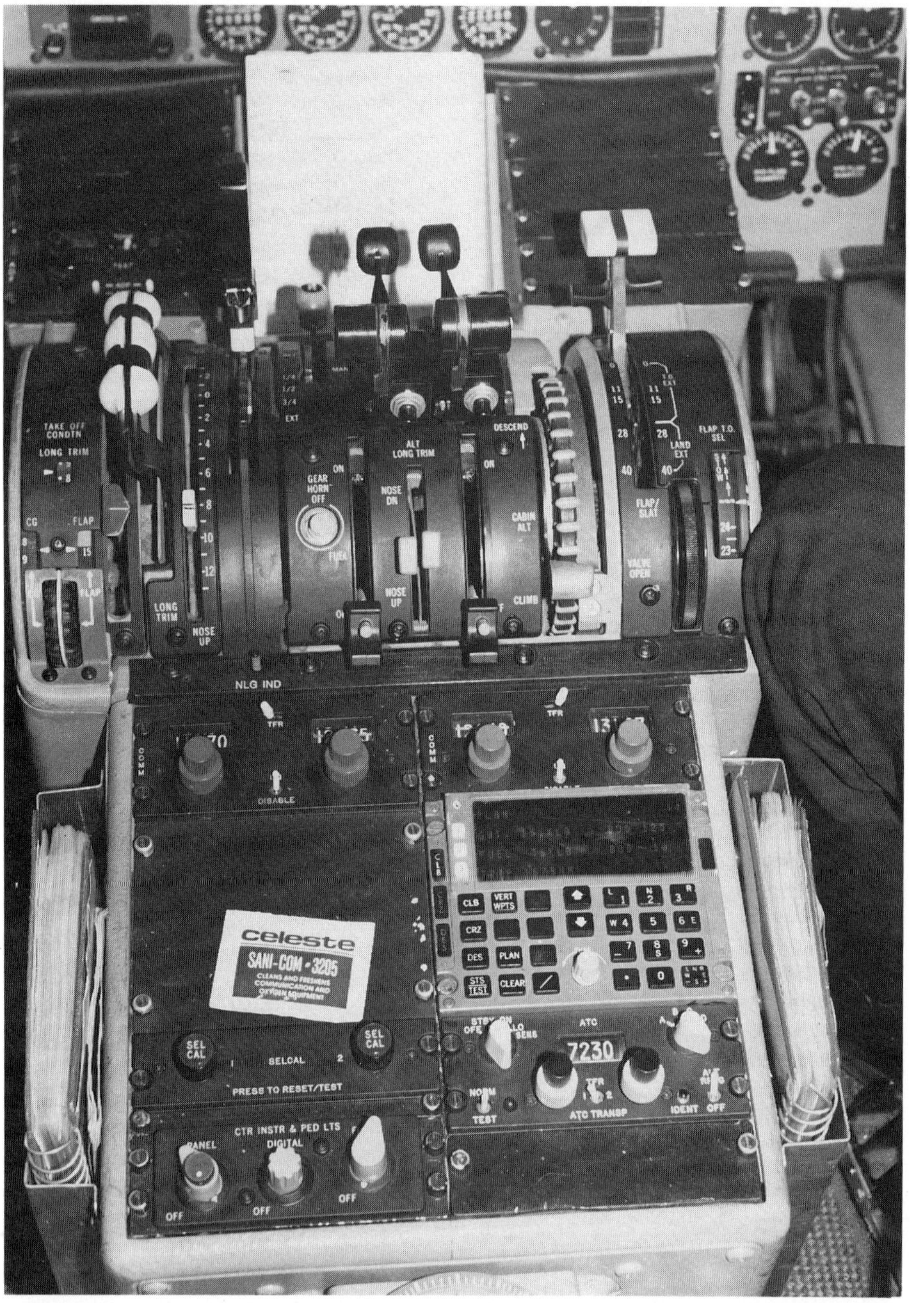

squarely at the booming inclusive tour market, Paramount had a successful first year and doubled its fleet in 1988 with two more MD-83s, operating out of Bristol, Birmingham, Newcastle and Belfast. Further aircraft, including Boeing 737s, are now on order for an expanded fleet in 1989.

Curious to sample this interesting aircraft, I was able to accompany Capt Dave Hipperson and First Officer 'Dusty' Rhodes on Paramount Flight PAT251 from Newcastle to Heraklion in Crete, and I joined them as they carried out their pre-flight checks.

The flightdeck of the MD-83 is superficially similar to most modern jet airliners but both pilots thought that it was well laid out with plenty of space, particularly outboard of the pilots' seats where maps, charts and the ship's library are stowed. An overhead panel comes right down to just above the top of the instrument panel coaming and houses the controls for the aircraft's ancillary systems. Between the pilots is a broad central console housing the throttles, fuel HP cocks, undercarriage, flaps and slat controls as well as communications equipment and the weather radar screen.

The aircraft was full with 165 passengers and, with 36,000lb of fuel on board, take-off weight was calculated at 151,776lb, the maximum allowable from Newcastle's 7,500ft runway in the prevailing conditions. Like many flights routeing on the busy airways to the eastern Mediterranean, PAT251 started the day with one hour's ATC delay before being cleared to start for an 07.45hrs departure slot time. As the MD-83 was pushed back by the tug, the crew started the engines — a straightforward process with the engines being spooled up to 30% N2 (N2 — core compressor rpm) before the fuel HP cock was selected 'ON'. With the engines stabilised at ground idle (30% N1, 58% N2 and exhaust turbine temperature at 350°C), the Captain waves the tug away and calls for taxy clearance. As the MD-83 trundles slowly down the runway before turning for take-off, the crew run through the pre-take-off checks and the cabin staff Number One reports the cabin secured. As the MD-83 swings through 180° to line up for departure, the effect of the unusually long fuselage becomes apparent. With the nosewheel set right forward,

Above left:
Engine, trim and flap controls together with communications equipment occupy the centrally positioned console.

Left:
On short final for Runway 27 at Heraklion. The threshold is at the top of a high cliff; at the far end of the runway is the built-up area of the town.

Above:
With the aircraft cruising at 29,000ft, the First Officer checks the fuel state and engine instrument readings. The airspeed indicator shows a speed of just under 300kt.

below the flightdeck, precise steering is easy and free of some of the problems which occur on aircraft with the nosewheel set well back (such as the Boeing 747) where a considerable overhang of the taxiway edges has to be allowed for when making turns. In a sharp turn such as lining up at the end of a runway, the flightdeck of the MD-83 actually moves sideways as the nosewheel castors through almost 90° — a strange sensation.

Another sensation, almost unique to this aircraft, is the complete absence of any engine noise as the throttles are opened and the aircraft accelerates for take-off. With the engines set right at the back of the large fuselage, the only sounds heard are the routine calls of the pilots, increasing aerodynamic noise and a rising hum from the twin-nosewheel bogie. The only indication of the power being delivered is from the engine instruments as the big-fanned JT8D-219s spool up to take-off power.

At our take-off weight the critical speeds have been calculated at 145kt V1, 151kt VR, and 155kt V2. These are respectively the stop/go decision speed, the point at which the aircraft is rotated for take-off, and the single-engined safety speed, and all are called out by the Captain as the First Officer performs the take-off. Once airborne the flaps are retracted as the MD-83 accelerates through 170kt and the leading edge

Above:
Taxying in to the crowded apron at Heraklion. Visible are a DC-10 of Calair and the Olympic Airways A300 which landed a few minutes ahead of the Paramount MD-83.

Below:
A fine view of the MD-83 on the ground at Heraklion as it turns in to its allocated parking bay.

slats are taken in at around 210kt. By this time the aircraft is climbing at 2,000ft/min and has turned left on to a heading of 134° towards the DOGGA intersection in the middle of the North Sea. Altitude is 4,500ft and ATC have cleared us initially to Flight Level 170 (FL170=17,000ft).

At 08.00hrs, 7min after take-off, we are passing FL120 with engine parameters set at 92% N1, 94% N2, EGT 500°C, and fuel flow around 7,000lb/hr per engine. Both autopilot and autothrottle are engaged. Cleared successively by Pennine Radar to FL260 and then by London Control (North Sea Sector) to FL290, the aircraft reaches 29,000ft at 08.10hrs, 17min after take-off. Power settings are reduced to give a cruising speed of 300kt IAS and each engine is burning 3,000lb of fuel/hr. This equates to a True Airspeed (TAS) of 451kt, translating to a ground speed of 488kt due to a strong tailwind (which will add over half an hour to the return flight later in the day).

Settled in the cruise, there is time to look at some of the aircraft's systems. Foremost amongst these is the Performance Management System (PMS) with its control panel and LED display set in the centre console. Integrated with the autothrottles, the PMS calculates and flies the aircraft through the optimum vertical flight profile. At the commencement of the flight the pilots enter the total track mileage and the PMS will then work out the best climb and descent profiles as well as the optimum cruising level at each stage of the route. It will then monitor and adjust the engine parameters through the autothrottle system to achieve the combination of airspeed, rate of climb and fuel burn best suited to the circumstances.

The PMS can also be used to give information to the pilots such as maximum and minimum safe speeds appropriate to aircraft configuration, weight and altitude. It will also give readouts of the distance run to Top of Descent (TOD) and Bottom of Descent (BOD). The latter is normally set at 10,000ft, 30 miles out from destination, and this information is extremely useful when requesting climbs and descents from ATC.

Another noticeable feature in the flightdeck is the digital readout fuel status display which, as well as showing total fuel and the remaining fuel in each tank, also displays the aircraft's gross weight as fuel is burnt off.

Although the aircraft's performance control uses the latest digital techniques, this particular aircraft (G-PATC) carried only the standard dual VOR/DME and ADF navigation equipment. This was backed up by a multi-mode autopilot with controls mounted in the glareshield above the instrument panel. Two of Paramount's aircraft are thus equipped, but the third and subsequent aircraft are fitted with the Omega area navigation system which offers considerably increased route flexibility and is independent of the ground-based VOR/DME beacons. Flight plan routeings can be fed into the system before departure allowing completely automatic navigation and relieving the pilots of many repetitive tasks.

Cruising over the North Sea and Holland at FL290, the MD-83 comes under the control of the Eurocontrol Air Traffic Control Centre (ATCC) at Maastricht. The next highest cruising level (offering more efficient fuel consumption) is FL330 but the PMS shows that at our present weight the aircraft will not be able to efficiently attain that level until 90min after take-off when nearly 10,000lb of fuel will have been used up. On transferring to Rhine Control (132.32MHz) at 08.49hrs the Captain advises ATC that he will be requesting climb to FL330 at 09.01hrs. At the appropriate time, clearance to climb is given, shortly after the MD-83 has passed overhead Frankfurt's busy Rhine-Main airport clearly visible below. By now the solid cloud layer which has covered Northern Europe is breaking up and the Alps present a spectacular vista in the distance.

The aircraft climbs steadily at 1,200ft/min until level at 330 and maintains this for the next two hours as it flies over southern Germany, across Austria and down the length of Yugoslavia. The crew perform various routine tasks — passing information to the passengers, gathering weather reports for various airfields along the route (Zagreb is shrouded in thick fog!), and checking out the fuel requirements for the return flight.

In the passenger cabin, a meal is being served from the aircraft's two galleys. One of these is comparatively spacious and is situated right forward while the other is on the port side about three-quarters of the way down the aircraft. The seating is five-abreast with a double row of seats on the port side and a triple row to starboard, arranged in 15 rows. Despite the length of the cabin, the styling of the overhead locker bays and the use of different surface materials avoids an obvious tube effect and the general impression was of a spacious and comfortable environment. Noise levels are very low and passenger reaction appeared very favourable — many preferring it to the six-abreast Boeing 737.

At 10.42hrs the aircraft entered Greek airspace with control being transferred from Belgrade to the Athens ATTC. Our route takes us over northern Greece and then down the country's eastern coast towards Athens. Twenty miles north of the city, ATC issues instructions for descent to FL290, although further descent is not given until crossing the island of Milos — only 90 miles from Heraklion and slightly later than the optimum point shown by the PMS. The crew have already carried out their descent checks while the Captain has calculated the landing weight at 124,000lb and has set out the performance card showing the approach and landing speeds at this weight.

By stages ATC clears PAT251 to descent to FL140 and control is transferred to Heraklion Approach

Below:
Passengers board for the return flight to Newcastle. Boarding is by means of the forward airstairs and the ventral stair beneath the tail.

(119.85MHz). The descent is made at 300kt, giving a rate of descent of 2,000ft/min and, with the engines throttled back, fuel burn is 1,000lb/hr per engine. Heraklion confirms that RW27 is in use and allocates an Expected Approach Time (EAT) of 11.38hrs. Our arrival time over the RKL VOR beacon will be 11.34hrs so a 4min hold is anticipated. A fresh northwesterly breeze is blowing and there is a broken cloud layer at 2,000ft.

Coming down to 5,000ft over the VOR, the First Officer puts the aircraft into a wide left-hand turn to lose the required four minutes as the Captain looks out to spot the aircraft landing ahead of us. This is an Olympic Airlines Airbus which we eventually see established on a long final for the runway but there is then a further delay as ATC clears an Olympic Dornier 228, which is already below us, to make an approach to RW31. Consequently it is 11.41hrs before the MD-83 is cleared to descend to 3,000ft and commence the approach procedure.

This requires the aircraft to fly away from the VOR on a track of 061° to a range of 10 miles before turning right to pick up an inbound track of 257° and continue descent to 1,015ft. As the First Officer turns on to the outbound track, speed is reduced to 200kt and the leading edge slats are extended. Turning inbound the speed has further reduced to 170kt and the landing gear is extended. The flaps are lowered in stages and speed is gradually reduced, aiming for a threshold speed of 129kt for landing with 40° of flap. With the aircraft established on the inbound leg, the runway is in sight ahead and the First Officer positions the aircraft visually on to the centreline.

Landing at Heraklion's coastal airport is tricky as the end of the runway juts out from a sheer cliff face falling away to the sea. The result is like flying over the round-own at the stern of an aircraft carrier with all the attendant problems of turbulence and height perception. Nevertheless we sweep in over the end of the runway, slightly fast at 136kt, and make a good touchdown at 11.48hrs. Reverse thrust is immediately selected and we slow down quickly, turning off to the left near the end of the runway.

The after-landing checks are quickly run through and the APU is switched on. Following ATC instructions the First Officer picks his way on to the crowded apron and the aircraft comes to rest at 11.51hrs — almost exactly four hours since leaving Newcastle. The aircraft has performed impeccably and after refuelling will be ready for the return trip, although in the event further ATC delays meant that the crew would not reach Newcastle until late in the evening — right at the end of the statutory 12¼hr maximum duty period.

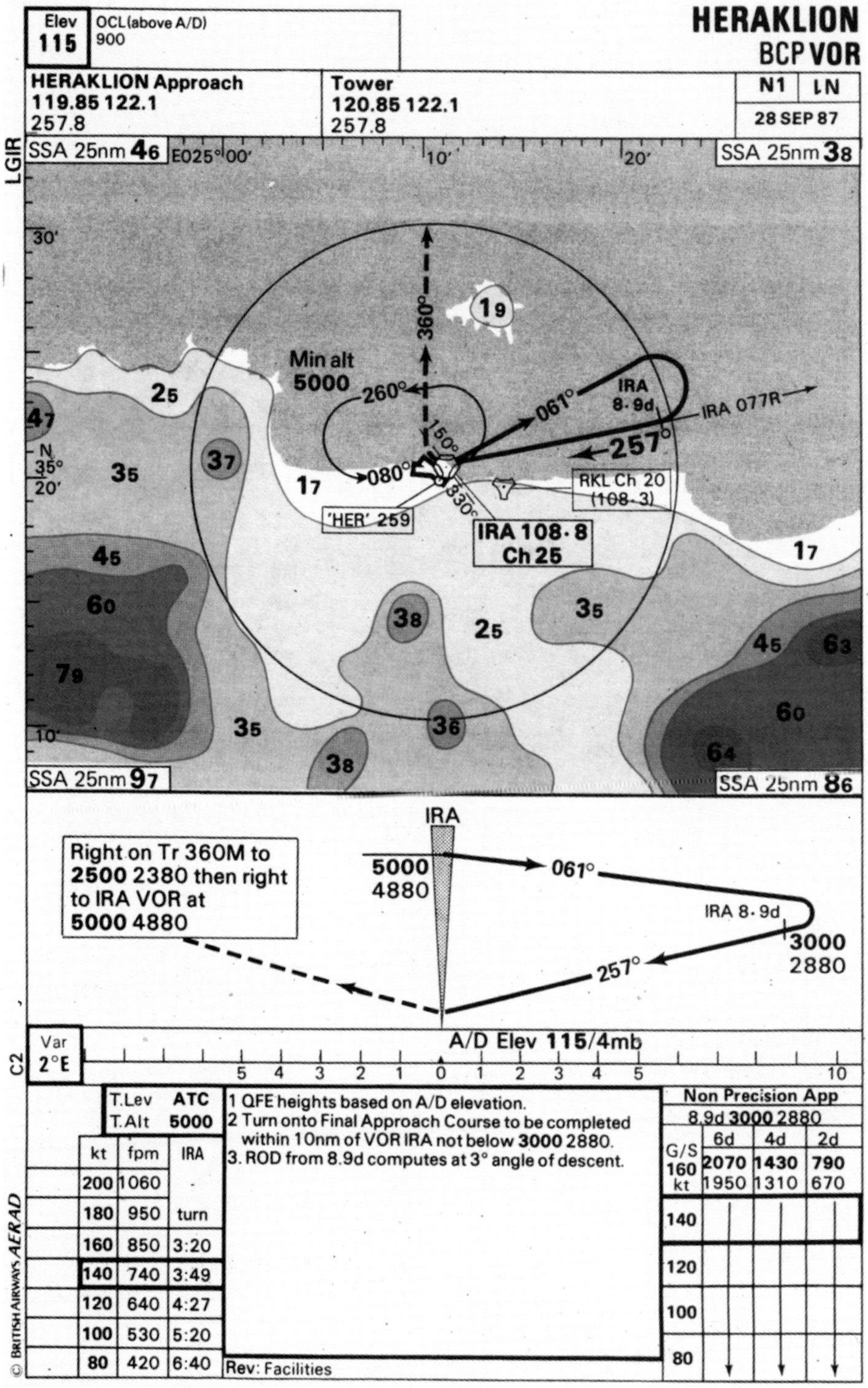

Above:

Chart showing the Cloudbreak procedure using the VOR/DME at Heraklion.

British Airways AERAD

(The author gratefully acknowledges the help and co-operation received from Paramount Airways in the preparation of this article. Particular thanks are due to Capt Hipperson and the crew of flight PAT251.)

Shorts — Pioneering to Privatisation

Today, such famous names as Avro, de Havilland and Supermarine have long since disappeared and now form only a part of the giant British Aerospace group. Up until now the Belfast-based Short Brothers can lay claim to being the world's first aircraft manufacturing company and one which has survived as a separate entity for 80 years. ***Leo Marriott*** *looks back over these momentous years, and also takes a look at what the future holds.*

THE BRITISH aircraft industry was born in the decade before World War 1 when men such as Verdon Roe, Sopwith, Rolls and de Havilland came to grips with the newly-discovered science and techniques of building heavier-than-air flying machines. These pioneers, and many others besides, founded companies and establishments able to produce thousands of aircraft, and which brought a new dimension to warfare from 1914 onwards. Today names such as Avro, de Havilland, Hawker and Supermarine have long since disappeared and form only a part of the history of the giant British Aerospace group while Rolls-Royce, risen phoenix-like from the financial disaster of 1971, is a new organisation which has absorbed all the other once-famous engine manufacturers.

However, until now there has been one exception to the process of takeovers and amalgamations — the Belfast-based Short Brothers which in fact lays claim to being the world's first aircraft manufacturing company, and one which has survived as a separate entity for 80 years.

The early history of the company echoes the experiences of many aeronautical engineers of the time.

Above:
A famous photograph showing the Short brothers (rear row: second, third and fourth on the right) and the Wright brothers (centre pair, seated) taken at Leysdown in 1909. *Shorts PLC*

AO·TEA·ROA
ZK-AMA

There were actually three Short brothers — Horace, Eustace and Oswald — who were born in the North of England between 1872 and 1883. It was Eustace who first became involved in aviation when he took up the new sport of ballooning in 1897, and in the following year he and Oswald started construction of their own craft at Hove in Sussex. By 1905 business had expanded enough to allow them to rent premises under the railway arches at Battersea where they built balloons ordered for both military and private use. Their activities brought them into close contact with members of the newly-formed Aero Club (later to become the Royal Aero Club) and it was this which led to their involvement in the construction of heavier-than-air machines.

The first orders, placed by Moore Brabazon and the Hon C. S. Rolls (famous pioneers in their own right), were for gliders and these were built between 1907 and 1909. There were also orders for powered machines and, recognising the need for more engineering expertise, the third brother (Horace) was recruited into the newly-formed company of Short Bros in 1908. At the same time activity on fixed-wing machines was moved to a new site at Leysdown on the Isle of Sheppey, although balloon work continued at Battersea for many years.

The first sustained flight in a powered aeroplane was made by the Wright brothers in America in 1903, and in 1908 Wilbur Wright travelled to Europe in order to demonstrate the brothers' latest machine. In a major coup, the Short brothers succeeded in obtaining approval to build the Wright machine under licence and a batch of six were laid down in 1909. This was the first quantity production order ever placed for any aeroplane and justifies the company's claim to be the first aircraft manufacturer in the world.

From this point the company expanded rapidly, moving to a larger site at nearby Eastchurch in 1910. This airfield played an important part in the early development of British aviation and was used by several pioneers as well as becoming the centre of naval flying prior to World War 1. It was this contact with the Admiralty that led to the company becoming involved in the production of seaplanes and flying boats. Perhaps the most famous was the Shorts Type 184 torpedo-carrying seaplane of which over 650 were built. It was employed in every theatre of action, including the Dardanelles, where one made the first successful aerial torpedo attack on a ship in August 1915.

The increase in naval activity at Eastchurch had led the company to move in 1914, this time to a riverside site at Rochester where completed marine aircraft could be launched directly into the River Medway. There was also a grass airfield just south of the town for landplanes. Later in the war, the company's airship and balloon activities were moved to Cardington in Bedfordshire where the massive sheds, still in use today by Airship Industries, were built for the 'R' series airships. This side of the company's business was taken over by the government in 1918 with only £40,000 compensation being paid for the company's assets — not the last time that Shorts received such treatment.

In the postwar years military orders virtually ceased although the company was able to gain experience in the construction of large flying-boat hulls by building a small number of F3 and F5 Admiralty-designed flying boats. However, in order to keep the factory active, Shorts turned to the production of motorboats, barges, and bodies for coaches, buses and trams. This work continued until the 1930s when the company's aeronautical fortunes began to take a turn for the better.

Although Shorts built several prototypes to various military speci-

Above left:
Between the wars the company made its name as a builder of large flying boats. The Shorts Sarafand, first flown in 1932, was powered by six Rolls-Royce Buzzard engines and was the largest British aircraft of its time, although only this single example was completed. *Shorts PLC*

Left:
Ao-Tea-Roa, one of two S30 'Empire' flying boats delivered to Tasman Empire Airways in 1939. *Shorts PLC*

Below:
Shorts' first commuter aircraft? The Scion carried five passengers and was powered by two 85hp Pobjoy Niagara engines. *Shorts PLC*

fications, it was civil aircraft which were eventually to lead to the biggest interwar orders. Based on experience with earlier types, the company produced the prototype Singapore in 1926. Unlike many of its contemporaries, this twin-engined biplane was of all-metal construction except for the fabric covering of the flying surfaces. Although military interest was cool — it was not until 1934 that the four-engined Singapore III was ordered in quantity for the RAF — the prototype was used by Sir Alan Cobham for a 23,000-mile survey flight around Africa in 1927.

Experience with the Singapore I placed Shorts in a good position to supply the newly-formed Imperial Airways with metal-hulled flying boats suitable for use on the developing air routes to the outposts of empire in Africa and the Far East. The result was the three-engined Calcutta which entered service in 1928. Later came the four-engined Kent which entered service in 1931 and was also produced as a landplane carrying up to 39 passengers. During the 1930s the company also produced the twin-engined Scion and a derivative, the four-engined Scion Senior, in small numbers — their first effort in the field of commuter aircraft!

The continuing success of Imperial Airways' empire routes, coupled to an indirect government subsidy in the form of substantial mail contracts, led to a requirement for a flying boat offering considerable improvements in range, speed and payload. Shorts proposals for a modern four-engined monoplane were immediately accepted and an order for 14 Shorts S23 aircraft was placed by Imperial Airways in 1934. This was later increased to 28 plus another three for the newly-formed Qantas Empire Airways. The 'C' class (or 'Empire' flying boats as they were referred to in service) put Shorts into the forefront of contemporary aircraft construction and when scheduled services started in 1936 the large and graceful boats

Below:
The flightdeck of Canopus, first of the 28 'C' Class boats ordered by Imperial Airways. Although spartan and simple by today's standards, it represented the peak of technology in the 1930s and bears more than a passing resemblance to the cockpits of the Sunderland and Stirling. *Shorts PLC*

Right:
The Sandringham was a simple conversion of the wartime Sunderland with the gun mountings faired over and a new interior. This example belonged to Norwegian Airlines, one of several postwar operators of the type. *Shorts PLC*

Below right:
Although this article concentrates on the civil aspects of Shorts' history, it should not be forgotten that the company has been heavily involved in the production of military aircraft. This is the scene at Belfast during World War 2 when the factory was committed to the production of Stirling bombers and Sunderland flying boats. This interesting photograph shows examples of the Stirling Mk IV and V, Sunderland Mk V, and the Percival Proctor (not a Shorts product). *Shorts PLC*

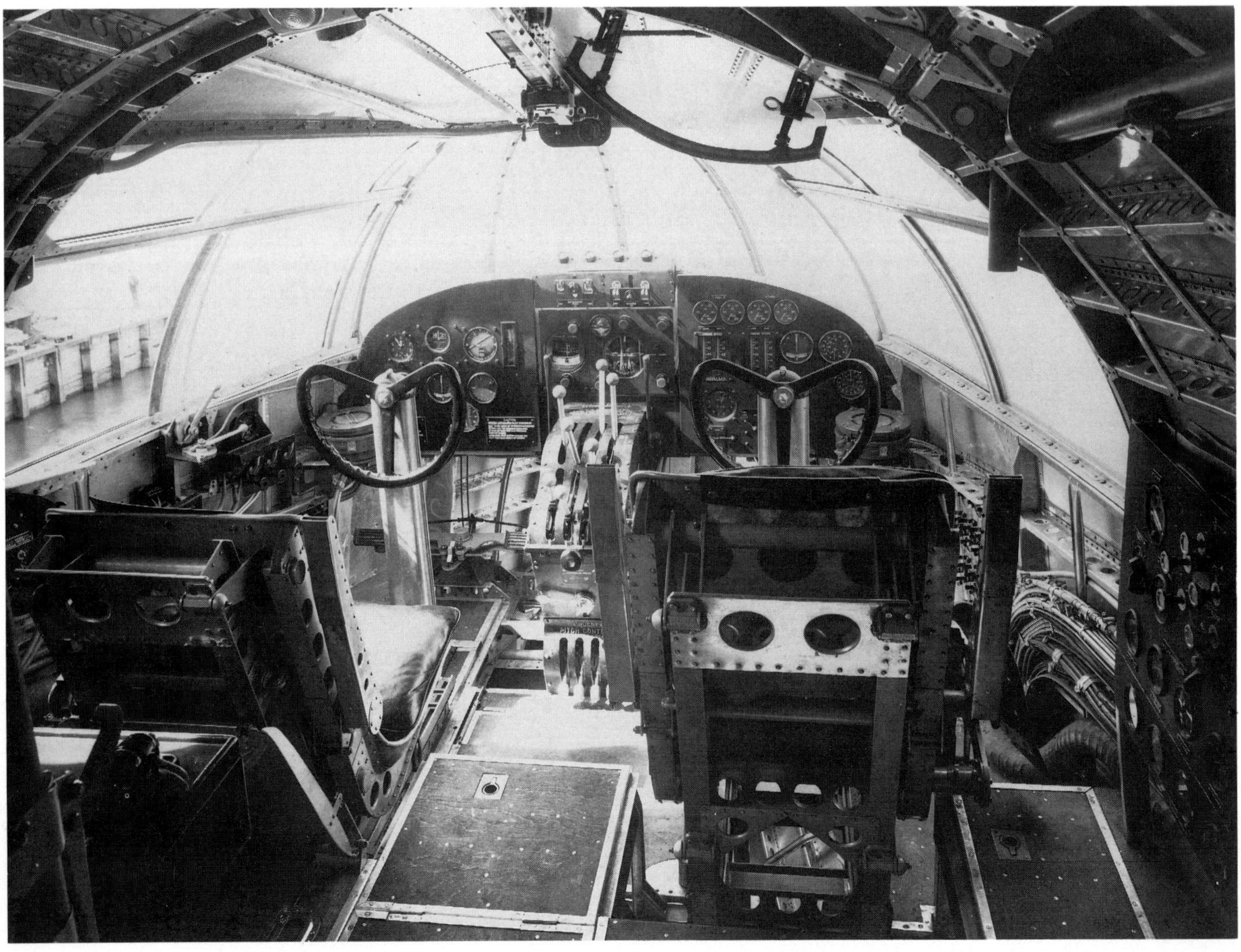

LN-IAW
DNL

became symbols of their age, in much the same way that Concorde is regarded today.

Following on from the 31 S23 boats were nine S30s with more powerful engines and increased range, and two S33s. The S30s were intended for trials on early transatlantic services and were among the first aircraft to be equipped for in-flight refuelling. Scheduled services commenced in August 1939 but were terminated at the outbreak of war. Final development of the 'Empire' class were the three S26 'G' class which first flew in 1939. These were larger than their predecessors and were powered by four 1,380hp Bristol Hercules engines. However, all three were requisitioned by the RAF and only one survived in commercial service at the end of the war. Mention should also be made of the *Maia-Mercury* composite aircraft (first flight 1937) which featured a small four-engined seaplane carried aloft on a modified 'Empire' boat and intended to give an early capability for transatlantic mail flights.

Shorts' war effort was centred on production of the famous Sunderland, of which over 700 were built in various versions, and the Stirling four-engined bomber (1,759 built). However, the war was not a happy period for the company. To cope with the build-up of work in the late 1930s a new factory was built at Belfast on the shores of Belfast Lough and run by a new subsidiary, Short & Harland. Initially this factory was employed in building Bristol Bombay transports and Handley Page Herefords under subcontract, but eventually it was given over to Stirling production while Rochester concentrated on the Sunderland. The stress of wartime production led to a number of difficulties which were experienced equally by other companies. The performance of the Stirling was inadequate (due almost entirely to a faulty specification) and enemy action had caused considerable disruption at Rochester.

By this time 60-year-old Oswald was the only survivor of the three brothers, and he was prevailed upon by the government to resign in 1943 so that nominees acceptable to the Ministry of Aircraft Production could be installed to run the company. Despite these actions, the government went further in March 1943 when it took over all the shares and assets of the company. To this day, the original shareholders and directors have never been compensated for this drastic action taken under the stress of wartime conditions.

Despite the political machinations, the work of the company went on but with the end of the war in sight a period of reorganisation was obviously called for. Despite the development of the S45 Seaford (originally Sunderland Mk IV) for use in the Pacific war, and the construction of two Shetland prototypes, military work quickly tailed off after 1945 and the company turned again to the civil market, its main customer now being BOAC.

The airline had employed 28 converted Sunderlands during the war and after 1945 these were quickly refurbished and renamed as

Left:
Developed from the Seaford, the Solent was produced in small quantities after the war for BOAC, but the airline ceased to operate flying boats in November 1950. *Shorts PLC*

Above:
The Short Sealand was an attractive amphibian which first flew in 1948. However, a crash during a Scandinavian sales tour the following year did not help sales and only 23 were sold. *Shorts PLC*

Below:
Over 150 Skyvans were sold and the aircraft laid the foundations of Shorts' present strong position in the commuter aircraft market. The example shown is one of three operated by the Government of Malawi. *Shorts PLC*

the Short Hythe. A number of Sunderlands were given a more substantial conversion involving streamlined fairings to replace the nose and tail gun turret positions, as well as a new interior seating 22 to 30 passengers on two decks. This version was known as the Sandringham and was ordered by Norway, Argentina and New Zealand as well as by BOAC. The first postwar production orders were for the Short Solent, derived from the Seaford, of which six were built in 1944-45 for the RAF. These were converted to Solents for BOAC and a further 16 were built for BOAC and Tasman Empire Airways.

However, the flying boat was already coming to the end of its commercial life as landplanes were now being introduced with improved performance and payload capability. With the end of Sandringham and Solent production the Rochester factory was closed down and all activities concentrated at Belfast. Once again the company found itself excluded from substantial military orders. Twenty-four

Sturgeon twin-engined target tugs were produced for the Navy in the late 1940s and a substantial order for the Seamew anti-submarine aircraft was cancelled in 1957 after 19 had been completed. The four-jet Sperrin bomber which flew in 1951 was not ordered into production as the subsequent V-bombers had a better performance. One bright spot was the Short Sealand amphibian of which 23 were built.

In order to keep the factory busy, Shorts acted as sub-contractor to other companies and built 130 Canberras and 33 Britannias during the 1950s. In addition, a number of research prototypes were built to support work in sections of the industry. These included the SB-5 of 1952 which tested the configuration of the Lightning fighter and the SC-1, Britain's first fixed-wing VTOL aircraft. It was also at this time that the company embarked on development of the highly successful Seacat missile and this led to the birth of the Missile Division which is one of the company's major revenue earners today.

It was in 1959 that Shorts began to consider ways of establishing itself back in the business of producing its own aircraft and this led to the setting up of a Light Aircraft Division to produce a rugged utility transport aircraft which it was hoped would find a wide market. The resulting SC-7 Skyvan was originally conceived as a simple airframe utilising a high-aspect ratio wing pioneered by the French engineer Hurel Dubois, powered by two Continental piston engines and carrying two tons of freight. It was quickly realised that turboprops offered improved performance and the prototype, which first flew in

Above left:
Breaking into the American market has been the key to the success of the 330 and 360 programmes. American Eagle is the name given to American Airlines' own commuter network which is made up of a number of small operators including Command Airways, Metro Airlines, and Simmons Airlines — all of which fly Shorts' products. *Shorts PLC*

Below left:
Wing sets for the Fokker 100 are produced at Belfast. The Aerostructures Division is an important contributor to the company's revenue and has enabled it to retain expertise in the building of large airframe structures and components. *Shorts PLC*

Above:
Short Bros will be involved on the design and production of the Canadair Regional Jet following the sale of the company to the Canadian Bombardier group. Although Shorts' own FJX project is now cancelled, the RJ has already been ordered by British Airways.

1963, was re-engined with Astazou turboprops. The Skyvan found a ready market and later versions were powered by Garrett TPE331-201 turboprops while a de luxe 19-passenger version was known as the Skyliner.

On the military side, 10 of the enormous Belfast freighters were ordered for the RAF, the prototype of which first flew in 1964. Several of these have since found their way on to the civil market after they were phased out of RAF service as a result of the defence cuts of the mid-1970s and are currently operated by Heavylift Cargo Airlines.

Although the Skyvan sold steadily, it became apparent that greater capacity, particularly in the passenger role, was required. Coupled with changes in FAA legislation which allowed small US airlines to operate larger aircraft, a demand for a 30-seat aircraft was identified and this resulted in the SD3-30 (subsequently the Shorts 330) which was basically a stretched and streamlined Skyvan with more powerful 1,198shp Pratt & Whitney Canada PT6A turboprops. First flown in 1974, the 330 was an immediate success and to date has sold nearly 200 (including the military version, the C23 Sherpa).

The success of the 330 led to demand for a larger version resulting in the Shorts 360 which offered an extra two seating rows to give a total of 36 passengers. A new swept tailfin replaced the previous twin-fin and rudder assembly; uprated engines were also fitted. Like its predecessors, the 360 has been ordered in quantity and over 150 have been built with orders still flowing in to match a current production rate of two aircraft a month.

Apart from producing its own range of aircraft, including the Tucano trainer for the RAF, Shorts has also built up a considerable volume of work in its Aerostructures Division. Work here includes building complete wing sets for the Fokker F28 and 100 jetliners; wing flaps, rudders and undercarriage doors for Boeing on their 737, 747 and 757 aircraft; and complete engine nacelles for the RB211, the IAE V2500 fitted to the A320 Airbus, and the Textron Lycoming ALF502R fitted to the BAe 146.

The retention of the skills needed to build major aircraft components would have stood Shorts well in their last independent project, the jet-powered FJX commuter airliner. This clean twin jet, designed to carry 44-48 passengers, received a favourable response from airlines when it was announced in 1987. However, a go ahead for the project depended crucially on finding a suitable partner in the aerospace industry to share the development costs. The situation was further complicated by the announcement, in June 1988, that the company would be sold off by the Government as part of its privatisation policies.

Several companies, including Boeing and Dornier, expressed an interest in buying Shorts but the final contenders were Bombardier of Canada (owners of the Canadair Co), and a consortium of Fokker and GEC. In June 1989 it was announced that the Canadian company was the successful buyer.

The takeover by Bombardier inevitably led to the cancellation of the FJX although Shorts will have a major role to play in the design and production of the 48-seater Canadair Regional Jet. A boost to this programme was given by an announcement that British Airways would be ordering at least 20 of the new aircraft. It is to be hoped that Shorts will prosper under its new owners but there will inevitably be regret at the changing status of Britain's oldest aircraft manufacturing company.

ATC Radar Systems

*Radar is the primary tool used by Air Traffic Control to guide aircraft safely through the overcrowded skies. Recent events have highlighted the congestion in Europe and the provision of new equipment is vital if the challenge of the 1990s is to be met. In this feature **Harry Cole**, Senior ATC Systems Consultant with Marconi Radar Systems Ltd, looks at the technical advances which will enable better use to be made of the limited airspace available.*

RADAR was one of the technologies which made tremendous advances during World War 2. As aircraft performance increased in leaps and bounds, radar techniques — the means to detect them — developed equally fast. Since 1945 continuing research and development has kept pace with still-increasing aircraft performance so that today's civil aircraft operate in a total radar environment in many parts of the world. In the period from 1950 until the mid-1970s it was usual to classify radars as in the accompanying table.

Until very recently, signal processing techniques did not allow one radar to perform adequately in any more than two overlapping roles (eg en route and TMA or TMA and Approach). The very few exceptions required massive antennae and were costly to build and maintain. As a

ATC RADAR EQUIPMENT CLASSIFICATION (1950-1975)

Role	*Typical range (nm)*
En Route Radar (ERR)	Up to 250
Terminal Area Radar (TMA)	Up to 180
Approach Control or Surveillance Radar (ACR/ASR)	Up to 60
Airfield Control Radar (ACR)	Up to 40
Airfield Surface Movement Indicator (ASMI or ASDE)	Up to 4

Above left:
Radar is an essential part of modern air traffic control systems. Typical of modern airport installations is this Plessey Watchman at Dubai airport.
Plessey Radar L246

Above:
Britain's Civil Aviation Authority has recently completed a major radar modernisation programme. This system at Heathrow is one of seven similar installations around the country which provide coverage for en route, TMA and approach functions. *Marconi Radar*

consequence most manufacturers developed a wide range of equipment which was a costly exercise in terms of research and marketing.

From the mid-1960s Secondary Surveillance Radar (SSR) started to become widely used as a result of the widespread fitting of SSR transponders in civil aircraft. In this system the transponder is interrogated by the ground radar and transmits information on the aircraft's identity and altitude which is then displayed on the radar screen. The success of this system led many ATC administrations to use SSR for the en route role, turning their back on more expensive long-range primary radar equivalents.

Modern signal processing performance has had a profound effect on primary radar development. The crude (by present-day standards) techniques and technology in early days made the 50cm wavelength most suitable for transmission as unwanted returning signals could be more readily dealt with at this frequency. As signal processing became more sophisticated, the shorter 23cm wavelength became popular for long-range radars and a 10cm wavelength is now almost standard for approach control radars. In addition, the reduction in the frequencies available for aeronautical use in the 600MHz (50cm) band has also had an effect on ERR radars.

As far as data handling and radar display technology is concerned, the current state-of-the-art has outstripped ATC procedures so that it is up to the ATC authorities to define their requirements, secure in the knowledge that the technology already exists to provide a solution. A typical example is the availability of sophisticated raster scan displays which use TV techniques to display radar-derived and other data using a variety of symbology, in full colour if required. At present very few ATC agencies have specified such equipment.

To review the current state of affairs it is appropriate to look first at en route radars where SRR is now

playing a major role. However, it is interesting to note that the UK CAA has just completed acceptance of the last of seven new all-purpose primary radars which can be used as ERRs at ranges of up to 256nm with the antenna rotating at 7.5rpm. Alternatively they can be used as approach radar up to a range of 80nm using a faster rotation of 15rpm (faster rates of scan gives more frequent data renewal — important when aircraft are manoeuvring as opposed to maintaining relatively steady courses). As in other similar installations, the associated SSR antenna (in this case a Marconi S1095) is mounted atop the primary radar reflector which is built by Telefunken AG and weighs some 20 tons, measures 14.5m across and is 9m high. As is now common technique, the reflector produces two separate beams which share a common azimuth boresight but have different elevation envelopes. This assists the tracking of aircraft targets through areas of ground clutter when the higher beam gives a better response. In this international installation, the radar electronics are produced by Hol-

Above:
The new radar facility at Jordan's Queen Alia airport exemplifies the modern strategy of using SSR out to 200 miles and primary radar to 80 miles.
Marconi Radar 8612012

Below:
The parabolic reflector of a Marconi S511 series Approach/TMA primary radar is designed to accept a monopulse SSR aerial above, as shown in this illustration.
Marconi Radar 8412123/30

landse Signaal Aparaten and the transmitter/receiver uses pulse compression techniques which reduce the transmitter power required to a peak of 150kW rather than the many megawatts which would have been needed with traditional designs.

In the field of TMA radars there are numerous manufacturers offering suitable products (Selenia of Italy and Thompson CSF of France to name but two, in addition to four British companies) and competition for orders is intense. Again, modern signal processing techniques have had an effect on the types of transmitters/receivers in use. In the late 1960s, klystrons replaced the simpler magnetron but recent developments in modulators and magnetrons have restored the balance.

Almost all the modern Approach Surveillance Radars (ASR) operate on the 10cm waveband and many give excellent performance out to a distance of 80nm. This has encouraged some ATC authorities to use them in the en route role with longer ranges being covered by SSR only. This is an extremely cost-effective solution to some requirements, and a typical example is the installation at Jordan's Queen Alia Airport. The economy of such a combination appeals to many of the world's smaller airports as it offers an effective method of increasing the amount of traffic which can be safely handled and thus increase revenue.

The American approach to modern ASR requirements has produced the Westinghouse ASR9 which is designed to meet stringent FAA specifications. It includes facilities to detect and outline weather returns without obscuring returns from aircraft targets. As a consequence, most other manufacturers now offer this facility.

The performance and relatively low cost of modern ASRs has reduced the need for a separate breed of Airfield Control Radars such as the Plessey ACR430, although equipment of this type continues to be used at many airports, mainly as a back-up to more modern installations.

One radar which has been developed to a remarkable degree of sophistication is the very short-ranged Airfield Surface Detection Equipments (ASDE) which use very short wavelengths (around 3cm) to give high resolution. Thus a typical ASDE (or ASMI) will show the shape of an aircraft or pick up a target as small as a pedestrian, or even a small animal anywhere in an airfield. Such equipment is vital at busy airports during periods of poor visibility. It is startling to realise that some aircraft can now land in visibilities of less than 100m, but pilots and controllers then have great difficulty in preventing accidents on the ground. This is where modern ASDE radars are so important.

Although SSR radar techniques date back to before the war, modern designs have almost totally solved the system's notoriously poor quality of track smoothness at medium and long ranges by the use of the monopulse technique. The reliability of SSR is inherently greater than primary radar as much lower power outputs are required. Data from SSR returns is normally

Below:
The LAR 200 Local Area Radar has been designed by Racal Avionics for use at smaller airports and features a full colour raster scan display which can be viewed in normal daylight. *Racal 22545/12*

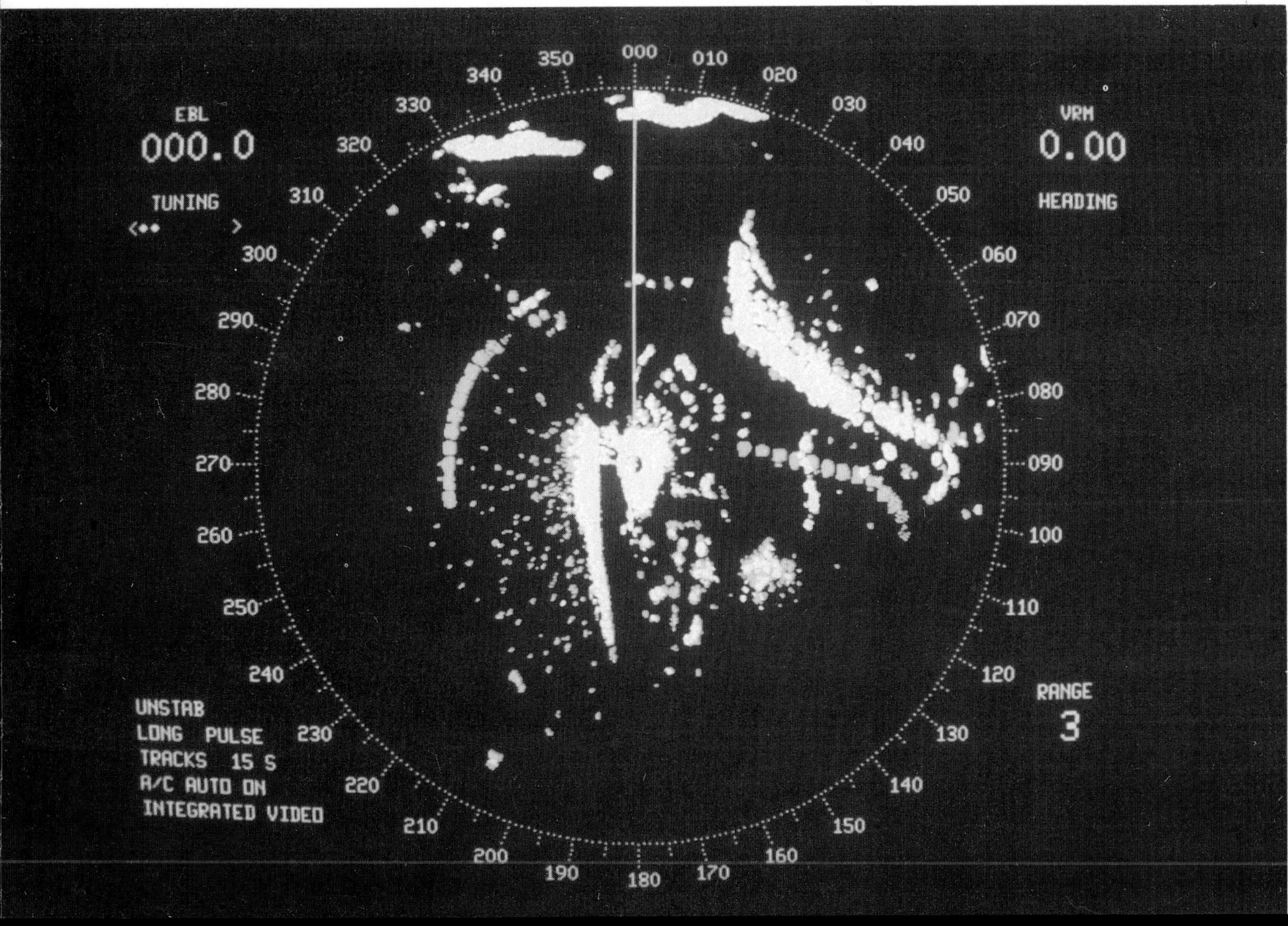

Above:
The fast-rotating antenna of a Racal 18X Airport Surface Movement Indicator being installed at Manchester Airport. *Racal*

Above right:
Modern ASMIs, such as the Racal 18X, give an incredibly detailed picture of the airport with its runways, taxiways and parking aprons. This installation is at Gatwick Airport. *Racal*

Right:
A typical alphanumeric data readout from a modern SSR. In this instance each aircraft 'blip' has a data block which shows the transponder code selected (top line) and present altitude. For example the aircraft at bottom left is at 5,100ft. More sophisticated data processing could result in the aircraft's callsign and other information being shown. *Plessey Radar K382/4*

processed by a radar data computer which extracts details of target position, identity, altitude and track history, and displays this on the radar screen in alphanumeric form.

An important factor in modern radars is the total cost of a system over its expected in-service life, and consequently great attention is paid to the ease of maintenance and fault repair. It is common to find monitoring of each line-replaceable unit (LRU) — such as a printed circuit board — so that relatively low-skilled staff can identify and replace a faulty unit. Higher technical skills can then be brought to bear at a main workshop to effect a full repair. Automatic test equipment (ATE) is frequently supplied to speed up the process of fault finding in complex miniaturised printed circuits. In systems using a number of radars scattered over a large area, central monitoring can be used — often under computer control. The UK CAA has implemented such a scheme for its very large radar and radio navigation aid infrastructure and this results in economies in spares holdings and technical skills required.

Future trends will shortly see the implementation of the SSR Mode 'S' (selective addressing) system whereby data may be received and sent from and to an aircraft. The use of this, together with satellite communications for voice and data transfer, will enable accurate position information to be obtained well outside the capabilities of existing radar systems and improve communications and control techniques in areas such as the North Atlantic where radar is not effective at present. Airborne and Ground Data Link Processors (A/GDLP) will determine the optimum route for the passing of data.

Combined, these advances mean that it will soon be possible to achieve a high level of automation in ATC, as well as the passing of other data such as weather reports or in-flight engine conditions to other interested agencies such as the Met Office and airline operation centres.

For radars, both primary and secondary, it is likely that the familiar rotating aerials will eventually disappear as electronically scanned antennas are developed for civil use (these already exist for military applications). Looking even further ahead, and beyond the limitations of pure radar techniques, is the possible use of Intelligent Knowledge Based Systems (IKBS). Using massive computing facilities it would be possible to build up a database relevant to ATC which would include in-flight aircraft data derived from radar and data links, route structures, system capacity, traffic flowrates, operating methods and many other items. Sophisticated programmes could then act on this data to aid controllers in deciding optimum flightpaths for the mix of aircraft under their control with appropriate instructions being passed automatically. In this scenario, the controller would become a systems manager in much the same way that pilots now manage the complex flightdeck systems while rarely actually flying the aircraft themselves.

Given the present and forecast trends in traffic density, air traffic controllers around the world will look with great interest on the technological advances which will eventually come to their assistance. Whatever developments occur, it can be certain that radar-derived techniques will continue to provide the vital basis on which safe air traffic control systems are founded.

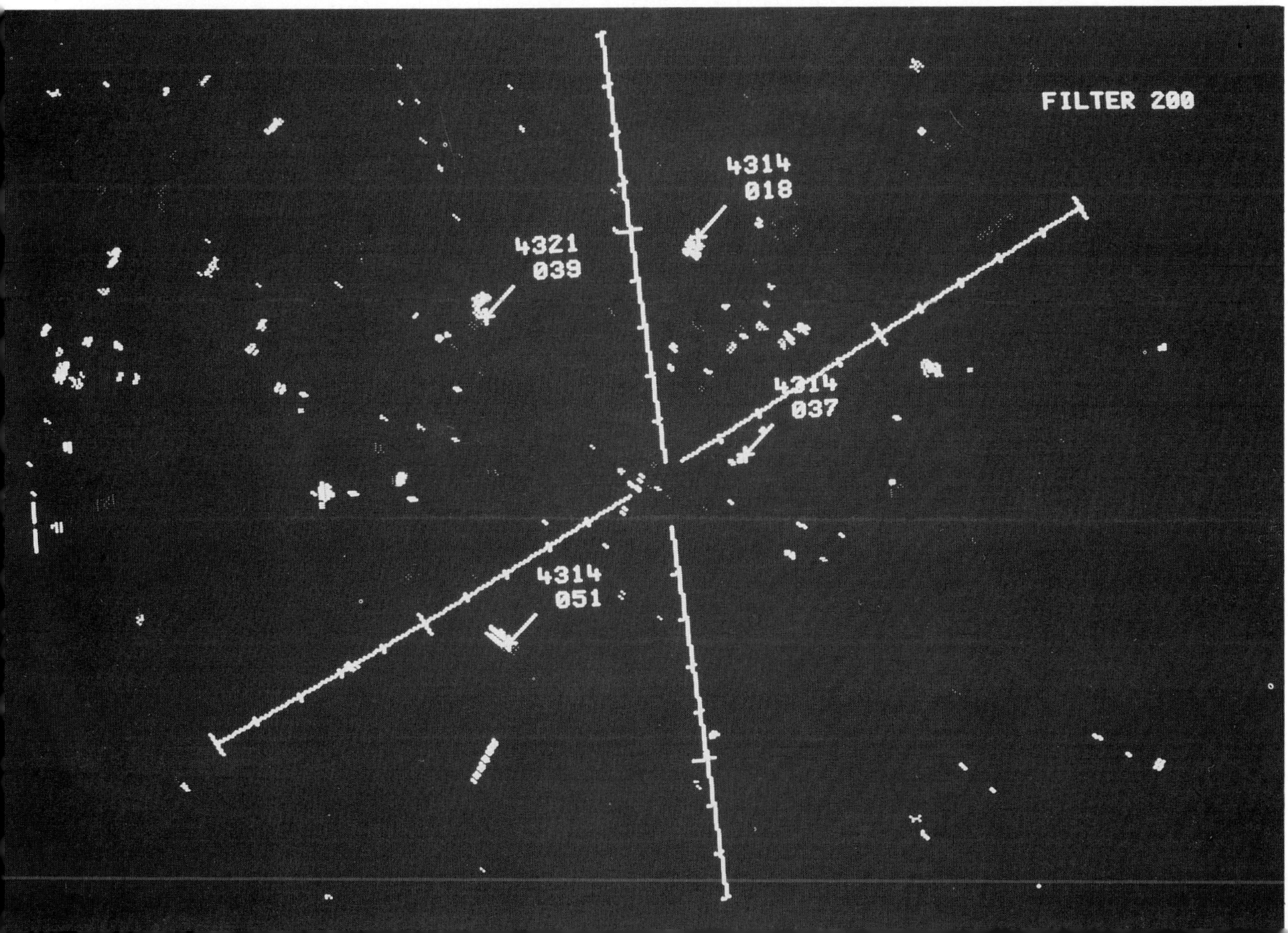
FILTER 200
4314
018
4321
039
4314
037
4314
051

Aer Lingus Commuter

***Leo Marriott** looks at the independently-managed subsidiary of Aer Lingus.*

'SHAMROCK three zero three, cleared via RUSH for ILS approach to runway two three. You are two seven miles from touchdown.'

Capt Tony Kearney acknowledges these instructions from Dublin Radar as he swings his aircraft on to a heading of 250° and adjusts the power settings to maintain a steady rate of descent. The atmosphere on the flightdeck is crisp and efficient as the crew run through the approach checks for the final stages of the flight from Newcastle in northeast England. Despite the 'big jet' feeling, this aircraft is in fact a Shorts 360 turboprop operated by Aer Lingus Commuter, an independently-managed subsidiary of the Irish flag-carrying airline.

As Flight EIN303 lands and taxies in, most of the other four Shorts 360s operated by the airline are already parked in front of the Dublin terminal. In most cases a slick 30min turnround will see these aircraft departing again for a variety of destinations within the Irish Republic or on the British mainland. As in other parts of the world, successful operation of commuter services depends on an efficient and dedicated ground organisation to keep the aircraft running punctually — and profitably. Indeed, it was the enthusiasm of Aer Lingus pilots and ground staff which was directly responsible for the birth of Aer Lingus Commuter in 1983.

Above:
Aer Lingus Commuter is a Dublin-based independent subsidiary of Ireland's national airline.

By the early 1980s Aer Lingus was an all-jet airline, having disposed of its Friendships and Viscounts, and running a fleet of BAC One-Elevens and Boeing 737s for its European and United Kingdom scheduled flights. The world trade recession at that time meant that several of the shorter routes could not generate enough passengers to support jet operations and in 1982 the Dublin-Belfast route was subcontracted to Avair, a small independent airline using a Shorts 330 aircraft. In the autumn of that year a staff meeting was called at which the Aer Lingus management explained that falling load factors meant that the airline would be abandoning several of its shorter routes (Dublin to Liverpool, Edinburgh and Exeter) and allowing Avair to take over the licences. The immediate reaction of the staff, much to the surprise of the management, was that if smaller airlines could operate these routes then surely Aer Lingus with all its resources could mount a similar operation.

Although this would mean reintroducing turboprops into the fleet, the crews were prepared to accept lower rates of pay to fly such aircraft. They reasoned that a more flexible career structure and earlier command prospects would eventually result while the retention of traditional Aer Lingus routes would only benefit the airline in the long run.

Given this response, it was decided that a trial operation would be mounted in 1983 using a single leased Shorts 330 and Aer Lingus Commuter was set up as a separate, but wholly-owned, subsidiary. Initially five pilots carried out conversion training on the 330 and flights to Liverpool and Leeds were started. Aircrew complement eventually rose to four captains, three first officers and six cabin staff.

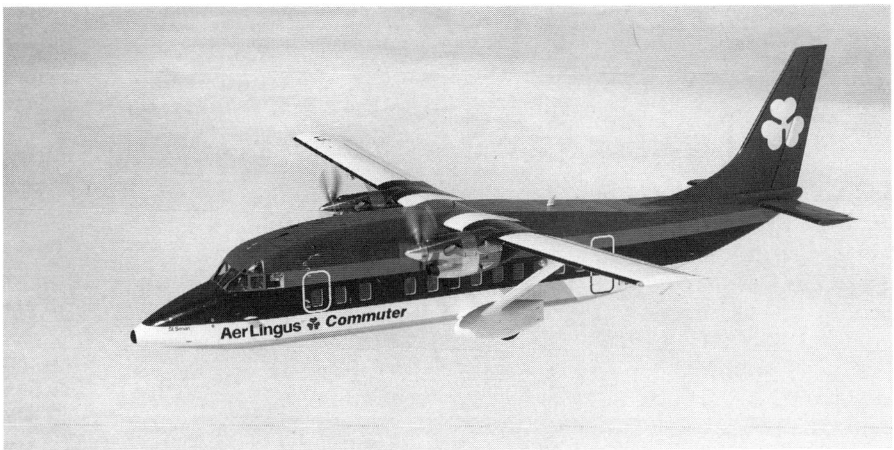

Above:
Backbone of the airline's fleet is the Shorts 360 of which five are currently in service. *Shorts PLC 84405-45*

Below:
Thanks to the ties between Aer Lingus Commuter and the parent airline, this young first officer can look forward to an interesting and varied career. After a period on the 360s he will move on to Boeing 737s, while his early command experience will bring him back to the Commuter fleet.

Following a very favourable passenger reaction to the new style of operation, two Shorts 360s were ordered at the end of 1983 and these subsequently entered service early in 1984 when the original 330 was retired. In the same year Avair, the independent airline which had triggered the Aer Lingus Commuter operation, ceased trading and most of its routes were taken over including Dublin to East Midlands. To cope with this new business and keep up with existing demand a further two 360s were ordered whilst a 330 on a short-term charter was brought in to help with the peak summer traffic.

A major setback occurred in early 1986 when EI-BEM, one of the 360s, crashed while on approach to East Midlands. This was caused by a combination of icing and wind shear but fortunately there were no fatal-

ities. Another aircraft (EI-BSP) was obtained as a replacement.

A fifth aircraft (EI-BVM) was added early in 1988 and new services were begun to Galway and Sligo in northwest Eire, and to Newcastle on the UK mainland. The latter service brought the airline into contact with Gill Aviation, a Newcastle-based airline then operating a fleet of three Shorts 330s on Datapost flights and *ad hoc* charters. With a requirement for further capacity in mind, arrangements were made to charter one of the 330s during the day when it was not required for overnight mail flights. The aircraft would position out to Dublin each day to operate a number of sectors before returning to Newcastle in the evening. It was quickly realised that the cost of the positioning flight could be reduced if passengers were carried, effectively doubling the number of flights available on the route. This was done and the results have been very pleasing for both airlines.

The routes to Galway and Sligo brought regular air services to these parts of the Republic for the first time, although many people expressed reservations as to the commercial viability of such flights. In fact they have proved very popular with two return flights a day to each, and a third on alternate days. Despite the operational difficulties posed by short runways surrounded by high ground and equipped with limited approach aids (NDB only), the number of diversions due to weather conditions has been well within acceptable limits. However, there are strict company criteria concerning the experience of crews rostered to this route.

The current route network includes flights from Dublin to Leeds, Liverpool, Edinburgh, Newcastle, Bristol and East Midlands, as well as domestic services to Cork, Shannon, Galway and Sligo. In the summer, weekend flights to Guernsey and Rennes cater for holiday traffic. While Dublin is the centre of operations there are also weekend

flights from Cork to Birmingham, Bristol, Manchester and Rennes.

To maintain this comprehensive network, each available aircraft is rostered for four return sectors each day, normally two in the morning and two in the afternoon and evening. In most cases these will originate from and return to, Dublin. Each crew will fly four single sectors in a duty period and the utilisation pattern allows this to be done on a shift system, covering either the morning or afternoon sectors.

As far as possible the airline is run as a separate entity from the parent company and employs around 95 staff under its Chief Executive, Ray

Above:
Shorts 360 flightdeck. *Shorts PLC 87258-2*

Left:
Three Aer Lingus Commuter Shorts 360s on the apron at Dublin preparing to operate the evening schedules. Passengers are currently bussed to their aircraft but improvements to terminal facilities will shortly mean direct access from the departure gates.

Wilson. This figure includes engineering staff presently operating from a collection of 'portakabins' clustered around the base of the old Dublin airport terminal building. Jokingly known as the Aer Lingus Commuter Village, these are being replaced by purpose-built premises, including a new hangar which will be built when present work on a new runway at the airport is completed. Despite the temporary premises, the engineering back-up is of a high standard with all aircraft stores listing and technical records being computer-based.

Although operationally independent, the Commuter organisation draws on many of the resources of Aer Lingus. All passenger handling, ticket sales, reservations and marketing is carried out by Aer Lingus. Ray Wilson freely admits that the Commuter airline's best marketing tool is the internationally respected Shamrock logo carried on the aircraft and featuring on all publicity material.

But if the smaller organisation benefits from being part of the Aer Lingus group, the reverse is also true. For a start, the Commuter services offer feeder routes to Dublin bringing passengers for other Aer Lingus flights. At a more practical level the existence of the Commuter airline allows an interesting exchange of aircrews with benefits to both parties in pilot training and experience. Newly-qualified pilots will join Aer Lingus as first officers on the Shorts 360 where they will gain considerable basic experience of airline operations at high workload levels. Most of the flights are short sectors in busy ATC environments and often in the worst of weather.

Eventually the pilot will transfer to the main Aer Lingus fleet as first officer on Boeing 737 or BAC One-Eleven jets. With this progression in mind, the flightdeck procedures and checklists have been carefully drawn up to be, as far as possible, representative of those used on the more advanced aircraft. It is this feature which accounts for the air of efficiency which is particularly noticeable on the flightdeck of the company's Shorts 360s. For example, it is standard practice on the jet fleet for the monitoring pilot to call out when there are 1,000 and 2,000ft to go to assigned altitude when climbing or descending. This procedure is also used on the 360 fleet but is carried out at 1,000 and

500ft to go, in order to reflect the lower performance of the turboprop.

After a period on jets, the pilot may return to the Commuter fleet as a captain, gaining early and useful command experience. Eventually, he may move back to the jet fleet to take a command on the larger aircraft. Thus both sections of the airline benefit from providing their pilots with the widest possible range of experience and training.

This cross-fertilisation of pilots, together with a steadily expanding fleet over the years, has meant that pilot training must be carefully integrated into the busy operating schedule. All training is achieved 'in-house' and for this purpose there is a Chief Training Captain supported by three other Training Captains and a Ground Instructor. Conversion courses are held in the winter and spring, and consist of two weeks' ground school followed by one week of flying training culminating in the base check. This is followed by a period of line training when a minimum of 16 sectors are flown to a representative selection of destinations before being line checked. For new captains this is followed by a further two weeks of command checking, again to a variety of destinations.

Apart from offering increased experience to pilots, the parent airline also benefits from the development of 'thin' routes by Aer Lingus Commuter. In fact the latter needs to be wary of its own success. As Ray Wilson comments, four 360 loads equate to one Boeing 737 and a time comes when it is seen to be advantageous to operate the larger aircraft. This is particularly relevant to the holiday traffic now being carried to the Channel Islands and northwest France. However the 360s will always be able to offer the flexibility of increased frequency, which is a vital selling point on routes aimed primarily at business traffic. Despite this, the volume of traffic and plans for longer routes led to evaluation of new aircraft for the 1990s. The frontrunners were the Fokker 50, the de Havilland Canada Dash-8 and the ATR42. All of these offered increased performance with a 50-seat payload but, as new designs employing up-to-date technology, also offered flexibility and economy of operation.

Final choice was the Fokker 50 and orders for two were placed in the latter half of 1988 and for a further pair in 1989. Due to the high profile of the airline and its association with a national flag carrier, an early delivery programme has been negotiated and the first aircraft should have been delivered in March 1989. These larger aircraft will considerably extend the airline's capacity on some routes.

The airline is now building on a history of steady expansion and success. No doubt the size and capacity of the route network will continue to grow, but it is to be hoped that the character of Aer Lingus Commuter will remain — basically a small and efficient airline run with a dedicated cheerfulness which is the hallmark of the Irish.

Aer Lingus Commuter

Fleet

A/C Reg	*A/C Type*	*Acquisition*	*Disposal*
EI-BEH	Shorts 330	04.83	11.84
EI-BEK	Shorts 360	03.84	in service
EI-BEL	Shorts 360	03.84	in service
EI-BEM	Shorts 360	06.84	01.86
EI-BPD	Shorts 360	11.84	in service
EI-BSP	Shorts 360-200	04.86	in service
EI-BVM	Shorts 360-200	03.88	in service

Utilisation (Revenue Block Hours)

Year	*Shorts 330*	*Shorts 360*
1983/84	2,139	—
1984/85	918	7,513
1985/86	—	8,773
1986/87	—	9,315
1987/88	—	10,087
1988/89	—	14,200

Pax Miles/Passengers

Year	*Pax Miles 000s*	*Passengers*
1983/84	5,879	36,106
1984/85	24,448	134,213
1985/86	27,948	150,008
1986/87	30,666	164,783
1987/88	35,724	195,785
1988/89 (budget)	37,777	209,271

Below:
The flightdeck of the Fokker F50 employs the latest technologies to reduce pilot workload. *Fokker 84/1/2412*